118

Advances in Biochemical Engineering/Biotechnology

Series Editor: T. Scheper

Advances in Biochemical Engineering/Biotechnology

Series Editor: T. Scheper

Recently Published and Forthcoming Volumes

Whole Cell Sensing Systems II
Volume Editors: Belkin, S., Gu, M.B.
Vol. 118, 2010

Whole Cell Sensing Systems I
Volume Editors: Belkin, S., Gu, M.B.
Vol. 117, 2010

Optical Sensor Systems in Biotechnology
Volume Editor: Rao, G.
Vol. 116, 2009

Disposable Bioreactors
Volume Editor: Eibl, R., Eibl, D.
Vol. 115, 2009

Engineering of Stem Cells
Volume Editor: Martin, U.
Vol. 114, 2009

Biotechnology in China I
From Bioreaction to Bioseparation and Bioremediation
Volume Editors: Zhong, J.J., Bai, F.-W., Zhang, W.
Vol. 113, 2009

Bioreactor Systems for Tissue Engineering
Volume Editors: Kasper, C., van Griensven, M., Poertner, R.
Vol. 112, 2008

Food Biotechnology
Volume Editors: Stahl, U., Donalies, U. E. B., Nevoigt, E.
Vol. 111, 2008

Protein – Protein Interaction
Volume Editors: Seitz, H., Werther, M.
Vol. 110, 2008

Biosensing for the 21st Century
Volume Editors: Renneberg, R., Lisdat, F.
Vol. 109, 2007

Biofuels
Volume Editor: Olsson, L.
Vol. 108, 2007

Green Gene Technology
Research in an Area of Social Conflict
Volume Editors: Fiechter, A., Sautter, C.
Vol. 107, 2007

White Biotechnology
Volume Editors: Ulber, R., Sell, D.
Vol. 105, 2007

Analytics of Protein-DNA Interactions
Volume Editor: Seitz, H.
Vol. 104, 2007

Tissue Engineering II
Basics of Tissue Engineering and Tissue Applications
Volume Editors: Lee, K., Kaplan, D.
Vol. 103, 2007

Tissue Engineering I
Scaffold Systems for Tissue Engineering
Volume Editors: Lee, K., Kaplan, D.
Vol. 102, 2006

Cell Culture Engineering
Volume Editor: Hu, W.-S.
Vol. 101, 2006

Biotechnology for the Future
Volume Editor: Nielsen, J.
Vol. 100, 2005

Gene Therapy and Gene Delivery Systems
Volume Editors: Schaffer, D.V., Zhou, W.
Vol. 99, 2005

Sterile Filtration
Volume Editor: Jornitz, M.W.
Vol. 98, 2006

Marine Biotechnology II
Volume Editors: Le Gal, Y., Ulber, R.
Vol. 97, 2005

Whole Cell Sensing Systems II

Applications

Volume Editors:
Shimshon Belkin · Man Bock Gu

With contributions by

H. A. Alhadrami · C. Baumstark-Khan · R. Boyanapalli · S. Buchinger · G. S. Bullerjahn · E. E. Diplock · Y. Fang · C. E. Hellweg · A. Hynninen · I. Karube · R. M. L. McKay · G. I. Paton · G. Reifferscheid · G. Reitz · S. Ripp · M. J. Rozmarynowycz · M. Shimomura-Shimizu · M. Virta

Springer

Editors

Prof. Dr. Shimshon Belkin
The Hebrew University of Jerusalem
Institute of Life Sciences
91904 Jerusalem
Israel
shimshon@vms.huji.ac.il

Prof. Dr. Man Bock Gu
Korea University
College of Life Sciences and Biotechnology
Amam-dong 5-ga
136-701 Seoul
Sungbuk-gu
Korea, Republic of (South Korea)
mbgu@korea.ac.kr

ISSN 0724-6145 e-ISSN 1616-8542
ISBN 978-3-642-26420-7 ISBN 978-3-642-12853-0 (eBook)
DOI 10.1007/978-3-642-12853-0
Springer Heidelberg Dordrecht London New York

Cover design: WMXDesign GmbH, Heidelberg, Germany

Printed on acid-free paper

Springer is part of Springer Science+Business Media (www.springer.com)

Series Editor

Prof. Dr. T. Scheper

Institute of Technical Chemistry
University of Hannover
Callinstraße 3
30167 Hannover, Germany
scheper@iftc.uni-hannover.de

Volume Editors

Prof. Dr. Shimshon Belkin

The Hebrew University of Jerusalem
Institute of Life Sciences
91904 Jerusalem
Israel
shimshon@vms.huji.ac.il

Prof. Dr. Man Bock Gu

Korea University
College of Life Sciences and Biotechnology
Amam-dong 5-ga
136-701 Seoul
Sungbuk-gu
Korea, Republic of (South Korea)
mbgu@korea.ac.kr

Editorial Board

Advances in Biochemical Engineering/ Biotechnology Also Available Electronically

Advances in Biochemical Engineering/Biotechnology is included in Springer's eBook package *Chemistry and Materials Science*. If a library does not opt for the whole package the book series may be bought on a subscription basis. Also, all back volumes are available electronically.

For all customers who have a standing order to the print version of *Advances in Biochemical Engineering/Biotechnology*, we offer the electronic version via SpringerLink free of charge.

If you do not have access, you can still view the table of contents of each volume and the abstract of each article by going to the SpringerLink homepage, clicking on "Chemistry and Materials Science," under Subject Collection, then "Book Series," under Content Type and finally by selecting *Advances in Biochemical Bioengineering/Biotechnology*

You will find information about the

- Editorial Board
- Aims and Scope
- Instructions for Authors
- Sample Contribution

at springer.com using the search function by typing in *Advances in Biochemical Engineering/Biotechnology*.

Color figures are published in full color in the electronic version on SpringerLink.

Aims and Scope

Advances in Biochemical Engineering/Biotechnology reviews actual trends in modern biotechnology.

Its aim is to cover all aspects of this interdisciplinary technology where knowledge, methods and expertise are required for chemistry, biochemistry, microbiology, genetics, chemical engineering and computer science.

Special volumes are dedicated to selected topics which focus on new biotechnological products and new processes for their synthesis and purification. They give the state-of-the-art of a topic in a comprehensive way thus being a valuable source for the next 3–5 years. It also discusses new discoveries and applications.

In general, special volumes are edited by well known guest editors. The series editor and publisher will however always be pleased to receive suggestions and supplementary information. Manuscripts are accepted in English.

In references *Advances in Biochemical Engineering/Biotechnology* is abbreviated as *Adv. Biochem. Engin./Biotechnol.* and is cited as a journal.

Special volumes are edited by well known guest editors who invite reputed authors for the review articles in their volumes.

Impact Factor in 2008: 2.569; Section "Biotechnology and Applied Microbiology": Rank 48 of 138

Attention all Users of the "Springer Handbook of Enzymes"

Information on this handbook can be found on the internet at springeronline.com

A complete list of all enzyme entries either as an alphabetical Name Index or as the EC-Number Index is available at the above mentioned URL. You can download and print them free of charge.

A complete list of all synonyms (more than 25,000 entries) used for the enzymes is available in print form (ISBN 3-540-41830-X).

Save 15%

We recommend a standing order for the series to ensure you automatically receive all volumes and all supplements and save 15% on the list price.

Preface

The last decade has witnessed an unprecedented convergence of biological, physical, chemical, and engineering sciences that allows the construction of integrated devices that could not have been feasible earlier. Diverse combinations of biotic entities with inanimate platforms are reported that repeatedly break new grounds in the engineering of biochips, biomimetic systems, and bioarrays. One exciting front in this continuously developing field deals the deposition and immobilization of live, functioning cells onto solid surfaces for biosensor applications. The present two volumes set attempts to summarize the state of the art in this field, to highlight several specific research aspects, to describe some of the most relevant applications, and to point out what we believe are the most important future directions for whole-cell sensor systems.

To accomplish this, leading scientific authorities on biosensor-related biological, chemical, and engineering aspects have joined forces by contributing 17 comprehensive review chapters that have been divided into two "Whole-Cell Sensor Systems" volumes. Volume I addresses the two main components of such systems: the cells on the one hand and the devices on the other; the second volume is devoted to a description of a set of present and future applications of whole-cell biosensors.

We have tried to direct the manner by which these issues are addressed here to illustrate the multidisciplinary nature that is essential for such an imaginative combination of diverse scientific disciplines. It is our hope that the resulting compendium of reviews will stimulate students, teachers, and researchers from all related fields to try and tread this exciting path.

Jerusalem
Seoul

Shimshon Belkin
Man Bock Gu
Editors

Contents

Adv Biochem Engin/Biotechnol (2010) 118: 1–30
DOI: 10.1007/10_2009_19

Published online: 20 January 2010

Applications of Microbial Cell Sensors

Mifumi Shimomura-Shimizu and Isao Karube

Abstract Since the first microbial cell sensor was studied by Karube et al. in 1977, many types of microbial cell sensors have been developed as analytical tools. The microbial cell sensor utilizes microbes as a sensing element and a transducer. The characteristics of microbial cell sensors as sensing devices are a complete contrast to those of enzyme sensors or immunosensors, which are highly specific for the substrates of interest, although the specificity of the microbial cell sensor has been improved by genetic modification of the microbe used as the sensing element. Microbial cell sensors have the advantages of tolerance to measuring conditions, a long lifetime, and good cost performance, and have the disadvantage of a long response time. In this review, applications of microbial cell sensors are summarized.

Keywords BOD sensor • Environmental monitoring • Food quality control • Medical diagnosis • Microbial cell sensor

Contents

M. Shimomura-Shimizu and I. Karube (✉)
School of Bioscience and Biotechnology, Tokyo University of Technology, 1404-1 Katakura, Hachioji, Tokyo 1920982, Japan
e-mail: karube@bs.teu.ac.jp

Abbreviations

BL	Bioluminescence
BMP	Bacterial magnetic particle
BOD	Biological (or biochemical) oxygen demand
DCIP	2,6-Dichlorophenolindophenol
DM	Double mediator
DO	Dissolved oxygen
FIA	Flow injection analysis
GGA	Glucose–glutamic acid
HCF	Hexacyanoferrate
HPLC	High-performance liquid chromatography
ISFET	Ion-sensitive field-effect transistor
JIS	Japan Industrial Standard
LAS	Linear alkylbenzene sulfonate
LD	Lethal dose
LOD	Lactate oxidase
MFC	Microbial fuel cell
OECD	Organization for Economic Cooperation and Development
OSS	OECD synthetic sewage
PD	Photodiode
PM	Photomultiplier
PNA	Peptide nucleic acid
RCI	Redox color indicator
RSD	Relative standard deviation
SP	Surface photovoltage
SPR	Surface plasmon resonance
TAS	Total assimilable sugar
TCE	Trichloroethylene

Microbes

Acetobacter aceti, *Achromobacter* sp. *Acinetobacter calcoaceticus*, *Arxula adeninivorans*, *Aspergillus niger*, *Aspergillus ustus*, *Bacillus licheniformis*, *Bacillus subtilis*, *Bacterium cadavers*, *Botrytis cinerea*, *Brevibacterium lactofermentum*, *Chlorella* sp., *Citrobacter freundii*, *Clostridium butyricum*, *Escherichia coli*,

Gluconobacter oxydans, Hansenula anomala, Issatchenkia orientalis, Klebsiella oxytoca, Kluyveromyces marxianus, Lactobacillus fermenti, Leuconostoc mesenteroides, Methylomonas flagellata, Moraxella sp., *Nitrobacter* sp., *Nitrosomonas europaea, Photobacterium phosphoreum, Pichia methanolica, Pseudomonas aeruginosa, Pseudomonas fluorescens, Pseudomonas putida, Pseudomonas stutzeri, Rhodococcus erythropolis, Saccharomyces cerevisiae, Salmonella typhimurium, Sarcina faecium, Sarcina flava, Scenedesmus* sp., *Serratia marcescens, Spirulina* sp., *Thiobacillus ferrooxidans, Thiobacillus thioparus, Torulopsis candida, Trichosporon brassicae, Trichosporon cutaneum*

1 Introduction

Since Karube reported in 1977 the first microbial cell sensor which used the whole cell as a biological sensing element [1], many kinds of microbial cell sensors have been developed as an analytical tool. Microbial sensors have been applied in important fields, such as environmental monitoring and food quality control [2]. Microbial sensors have replaced conventional methods, which are often complicated, time-consuming, expensive, and require pretreatment or clean-up of real samples prior to analysis. A few examples of microbial sensors in practical use in major fields are described in the following sections. Microbial sensors can be categorized by three applications, i.e., environmental monitoring (see Tables 1–5), food quality control (see Table 6), and medical diagnosis (see Table 7).

2 Environmental Monitoring

For environmental monitoring, microbial sensors have been most extensively studied and reviewed in the literature [3–7]. In recent years, biological (or biochemical) oxygen demand (BOD) sensors and toxicity sensors have been actively developed.

2.1 BOD Sensors

In environmental monitoring, the most important application is BOD determination in polluted water or effluent. In this review, the BOD sensors are categorized into several types, i.e., BOD_{DO} as dissolved oxygen (DO) type, BOD_{MF} as microbial fuel cell (MFC) type, BOD_{PL} as photoluminescence type, BOD_{MD} as mediator type, BOD_{DM} as double mediator (DM) type, BOD_{RC} as redox color indicator (RCI) type, and BOD_{SP} as surface photovoltage (SP) type. Table 1 shows a comparison of the biodegradation characteristics of organic compounds with these BOD sensors.

Table 1 Comparison of BOD values of various organic samples[a]

Substrate[b]	BOD_{DO}	BOD_{PL}	BOD_{MD}	BOD_{SP}	BOD_5
Glucose	0.72	0.62	1.54	0.66	0.50–0.78
Fructose	0.54	0.57	0.35	0.73	0.71
Sucrose	0.36	0.50	0.07	0.45	0.49–0.76
Lactose	0.06	0.31	0.02	0.04	0.45–0.72
Soluble starch	0.07	0.02	–	0.07	0.22–0.71
Asparagine	–	0.48	0.29	–	0.58
Alanine	–	–	0.73	–	0.55
Glycine	0.45	0.50	–	0.36	0.52–0.55
Glutamic acid	0.70	0.73	0.59	0.40	0.64
Histamine	0.35	–	0.27	0.34	0.55
Acetic acid	1.77	0.27	0.32	0.39	0.34–0.88
Citric acid	0.72	–	–	0.18	0.63–0.88
Lactic acid	0.17	0.32	0.66	0.14	0.40
Ethanol	2.90	0.25	0.30	0.49	0.93–1.67
Propanol	0.28	0.28	0.29	–	0.47–1.50
Glycerol	0.51	0.53	0.05	0.44	0.62–0.83
Reference	[11]	[62]	[77]	[91]	[12]

[a]Values are expressed in mg O_2 mg^{-1} substrate
[b]Concentration of each pure organic substances were 100 mg L^{-1} for glucose, 500 mg L^{-1} for sucrose, lactose, and glycerol, and 200 mg L^{-1} for others

2.1.1 BOD_{DO} Sensors

The first microbial BOD sensor applying the principle of MFCs was developed by our group [1] (see Table 2(a)). Anaerobe *Clostridium butyricum* was isolated from soil according to the Japan Industrial Standard (JIS) method. The microorganisms were immobilized by polyacrylamide gel and fixed to a glass cell. This sensor system realized a short measuring time of around 30 min. The measuring time was dramatically shortened by the invention of this BOD sensor compared with the conventional 5-day BOD (BOD_5) method. Thus, the possibility of BOD monitoring of effluent was suggested by this sensor development.

Subsequently, several microbial BOD sensors employing the batch and microbial membrane system have been developed by utilizing different immobilization methods with different aerobes microorganisms [8, 9] (see Table 2(b–d)). Microorganisms from soil were immobilized in collagen gel membrane, and anaerobes *C. butyricum* IFO 3847 were immobilized in a polyacrylamide gel membrane [8]. This system differs from typical batch systems with the existence of an air supply system. *C. butyricum* was grown and immobilized under anaerobic conditions, and the sensor system using *C. butyricum* was modified as an MFC. In this case, collagen gel was degraded by the soil microbes. Thus, the sensor system using *C. butyricum* subsequently improved, becoming easier to operate [9]. However, an anaerobe is not suitable for BOD estimation. Therefore, other approaches for microbial BOD sensors were required.

A flow system utilizing indication of respiration activity by an aerobe was subsequently developed for BOD_{DO} estimation [10] (see Table 2(e)). Soil

Table 2 Characteristics of early developed BOD sensors by our group

Sensor type	Microbe	Indicator/ transducer	Measurement range ($mg\ L^{-1}$)	Ref.
(a) BOD_{DO}	Soil consortium	DO/electrode	200–600	[1]
(b) BOD_{DO}	Soil consortium	DO/electrode	5.0–22	[8]
(c) BOD_{MF}	*C. butyricum* IFO 3847 (anaerobe)	DO/fuel cell	30–300	[8]
(d) BOD_{MF}	*C. butyricum* IFO 3847	DO/fuel cell	6–400	[9]
(e) BOD_{DO}	Soil consortium	DO/electrode	44–132	[10]
(f) BOD_{DO}	*T. cutaneum* IFO 10466, AJ 4816	DO/electrode	10–40	[11]
(g) BOD_{DO}	*T. cutaneum*	DO/electrode	2.0–20	[11]

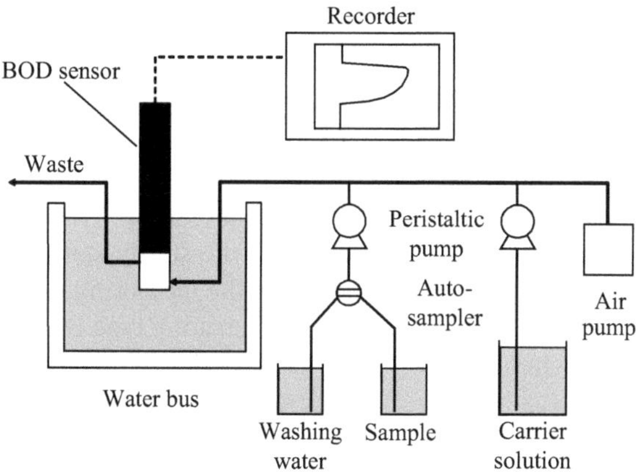

Fig. 1 Construction of a flow and porous membrane type BOD sensor

microorganisms were used and immobilized in collagen gel membrane as well as in the previous study [8]. Good reproducible results of 3% relative standard deviation (RSD) were obtained by the glucose–glutamic acid (GGA) standard solution, although the stability of the gel membrane was not enough and the same as in the previous study. Then the problem was raised as to which microorganisms of the soil consortium or single strain should be used for the BOD_{DO} sensor. For example, the biodegradability of the consortium to many kinds of organic compounds is greater than that of a single strain. However, reproducible results cannot be obtained by the use of a consortium; obtaining reproducible culture conditions in a consortium is almost impossible.

Thus, we next examined the use of a single strain and fabrication of a continuous flow system for automatic BOD_{DO} estimation [11] (see Table 2(f)). *Trichosporon cutaneum* AJ 4816 (IFO 10466) was employed as a single strain for a BOD sensor (see Fig. 1). Comparison of biodegradation characteristics of organic compounds between the BOD_{DO} sensor method using *T. cutaneum* AJ 4816 [11] was performed using the conventional BOD_5 method [12]. The sensor showed low BOD values compared with the BOD_5 method when soluble starch and lactose were employed for experiments. This might be caused from the slow decomposition rate of these compounds by the immobilized yeasts. On the other hand, the sensor showed high

BOD values compared with the BOD_5 method when ethyl alcohol and acetic acid were employed. These results suggested the oxidation rate of ethyl alcohol and acetate to be faster than that of some standard substrates such as GGA. Therefore, the GGA solution was defined as a standard solution for the BOD estimation of effluents. Before development of the BOD sensor, such primary effluent of wastewater or sewage from the pulp or food industry was hard to control by BOD values obtained by the BOD_5 method. As a new method for the BOD determination of primary effluent, a sensor method was established which was defined as a JIS (JIS K 3602) in 1990 [13]. Finally, an automatic flow system was developed for practical use [11], and about 800 instruments have been sold since 1983 [14] (see Table 2(g)).

Yang et al. developed disposable DO electrode chips that can be applied to the BOD estimation for field monitoring. At first the single DO electrode was constructed on silicon substrates using micromachining techniques [15] (see Table 3 (IIa)). This electrode is of the Clark type and *T. cutaneum* was directly immobilized on the electrode surface using an ultraviolet cross-linking resin. This DO electrode chip enabled measurements between 1.0 and 18 mg O_2 L^{-1} BOD. For BOD estimation using the DO electrode chip, a dynamic transient measuring method was adopted and compared with the steady-state measuring method as a conventional method [16]. In the study, the measuring time was dramatically reduced.

Subsequently, they fabricated an array type DO electrode (Clark type) using thin film technology [17] (see Table 3(IIb)). One array chip is formed with five DO

Table 3 Characteristics of BOD sensors recently developed by our group

Sensor type	Microbe	Indicator/transducer	Measurement range ($mg\ L^{-1}$)	Ref.
(I) BOD_{DO}	Thermophilic bacteria (not identified)	DO/electrode	1.0–10	[50]
(IIa) BOD_{DO}	*T. cutaneum* IFO 10466	DO/chip electrode	1.0–18 (det. lim. 0.2)	[15]
(IIb) BOD_{DO}	*T. cutaneum* IFO 10466	DO/chip electrode (five electrode array)	8.0–32	[17]
(IIIa) BOD_{DO}	*P. putida* SG10	DO/electrode	0.2–10	[18]
(IIIb) BOD_{DO}	*P. putida* SG10	DO/electrode	0.5–10	[19]
(IIIc) BOD_{DO}	*P. putida* SG10	DO/electrode	1.0–10	[20]
(IV) BOD_{DO}	*P. putida* SG10	DO/optical fiber probe (fluorescence)	1.0–10	[21]
(V) BOD_{PL}	*P. phosphoreum* IFO 13896	Luminescence/PD	20–160	[62]
(VIa) BOD_{MD}	*P. fluorescens* biovar V	Mediator/chip electrode	15–200	[77]
(VIb) BOD_{MD}	*P. fluorescens* biovar V	Mediator/chip electrode	15–260	[79]
(VIc) BOD_{MD}	*P. fluorescens* biovar V	Mediator/chip electrode (ten electrode system)	10–120[a], 25–250[b]	[82]
(VII) BOD_{RC}	*P. fluorescens* biovar V	Color indicator/PD	50–3000	[88, 89]
(VIII) BOD_{SP}	*T. cutaneum* IFO 10466 (AJ 4816)	pH/SP	10–100	[91]

[a]Sludge extract solutions (SDS)
[b]OECD

electrodes. However, these BOD_{DO} sensor chips developed for field monitoring have several remaining problems. The biggest problem is that the concentration of DO is limited to about 8 mg O_2 L^{-1}. Therefore, determination of high BOD samples influenced to the DO concentration.

Quality monitoring of secondary effluents (less than 10 mg O_2 L^{-1} BOD) is required for public organizations such as sewage plants and industries. In addition, the sensitive BOD_{DO} sensor was also required for monitoring low BOD values in river waters. Chee et al. developed pretreatment methods for the measured samples and highly sensitive BOD_{DO} sensors [18–23] (see Table 3(III) and (IV)).

Secondary effluents and the upper stream of river water generally indicate low BOD values, and they mostly contain refractory organic compounds such as gum arabic, humic acid, lignin, tannic acid, and surfactants [24, 25]. Thus, for the precise estimation of such samples, several microbes from sewage plants were screened and *Pseudomonas putida* SG10 was isolated to apply to highly sensitive BOD_{DO} sensors.

In the first step of the development of highly sensitive BOD_{DO} sensors using DO electrodes, a basic batch system was studied and characterized to obtain the optimum conditions for low BOD estimation [19]. The calibration curve was obtained by the GGA solution from 0.5 to 10 mg O_2 L^{-1} BOD and the detection limit was 0.5 mg O_2 L^{-1} BOD ($n = 5$). Next, in the BOD determination of various river waters (14 samples), the sensor's results generally indicated somewhat lower values than those obtained by the BOD_5 method. The reason could be that the compounds were not easily assimilable to the sensor in such a short time.

In the second step, a highly sensitive BOD_{DO} estimation was studied using an optical fiber probe as a DO sensor, based on fluorescence quenching by oxygen [18]. However, the reproducibility of this optical fiber biosensor was lower than that of the previously developed sensors. Therefore, the DO electrode was used again for subsequent studies.

In the third step, ozonation as a pretreatment method for low BOD samples was examined [20]. Ozonation is known as a treatment method to decompose organic compounds, especially refractory organics in industrial wastewater, municipal effluent, and river water. By pretreatment, humic acid and tannic acid were removed up to 22 and 18% respectively but other compounds were not removed effectively. In this study, ozonation of river water samples was successfully established as a new pretreatment method for low BOD_{DO} estimation combined with the sensitive sensor method. However, this sensor method requires ozonation equipment, which is expensive and large, and ozone gas requires careful handling. Therefore, a photocatalytic pretreatment method was next applied with the same aim.

Samples containing refractory organics can also be decomposed to biodegradable substances by photocatalytic preoxidation using illuminated titanium dioxide (TiO_2). Accordingly, the photocatalytic pretreatment method was applied to the sensitive BOD sensor [21]. The sensor responses to various river water samples, which were pretreated by photocatalysis, were successfully improved.

The results obtained by this sensor method and the conventional BOD_5 method corresponded. As the next steps, both a flow type sensor system and stopped-flow system based on the above-mentioned studies have been developed [22, 23]. As another highly sensitive BOD sensor, an MFC was applied using an improved cathode reaction [24].

Other BOD_{DO} sensors were also developed using different kinds of microbes, for example *Arxula adeninivorans* (salt-tolerant yeast) [26–28], *Bacillus subtilis* [29], *Hansenula anomala* [30], *Klebsiella* sp. [31], *K. oxytoca* [32], *Serratia marcescens* LSY 4 [33], *Torulopsis candida* [34], combinations of two microbes such as *Bacillus licheniformis* and *B. subtilis* [35–38], *Issatchenkia orientalis* and *Rhodococcus erythropolis* [39], *B. subtilis* and *T. cutaneum* [40], activated sludge [41, 42], a mixture of microorganisms (consortium) [43–45], MFC [46, 47], slime mold [48, 49], thermophilic bacteria [50] (see Table 3(I)), and dead bacterial sensor cells [51–55]. As other approaches, a respirographic BOD sensor was developed utilizing indication of CO_2 concentration (BOD_{CO2}) [56]. In addition, a disposable BOD sensor for measuring nitrification (N-BOD) and inhibition of nitrification in wastewater was developed using nitrifying bacteria [57].

2.1.2 BOD_{MF} Sensors

The first BOD biosensor, which was developed by Karube et al. in 1977, was based on an MFC using the hydrogen produced by anaerobe *C. butyricum* immobilized on the electrode [1]. Recently, applications of the mediatorless MFC to BOD sensor development have been performed using an electrochemically active metal-reducing bacterium [58–61]. The BOD_{MF} sensors can be used as continuous monitoring systems and have long-term stability.

2.1.3 BOD_{PL} Sensors

Using luminous bacteria (*Photobacterium phosphoreum* IFO 13896), a photoluminescence BOD sensor has been developed by Hyun et al. [62] (see Table 3(V)). The luminous bacteria, which are isolated from marine sources, emit light at a constant rate for fairly long periods of time as a result of normal metabolic processes. In this process, the substrate produced by organic compounds or catabolic degradation in the bacterial cell are shunted to the bioluminescence (BL) reaction which is coupled to the electron transport pathway [63]. In the experiment, *P. phosphoreum* was put into a transparent glass dish, which was located on a photodiode (PD) in a light-protected box. Signals from the PD were amplified and recorded. The BOD responses of the bacterial reagents could be observed between 20 and 160 mg O_2 L^{-1} BOD within 15 min with 7%RSD. Through research efforts (Tamiya et al.), the BOD_{PL} sensor was produced by Ishikawa Seisakusho Ltd. In addition, the sensor system was applied to an array chip for on-site BOD sensing [64]. This BOD_{PL} sensor will be applied widely to the field of environmental analysis by adoption to Japan Industrial Standard.

2.1.4 BOD_{MD} Sensors

The mediator has been applied to the fabrication of MFCs [65, 66] and to bacterial detection [67, 68]. It has been suggested that reduction of the RCI, rather than DO, is due to metabolic reactions of microorganisms [69]. Thus, instead of DO, hexacyanoferrate (HCF(III)) has been used as an electroactive compound for the development of amperometric biosensors using microorganisms [70, 71].

The first BOD_{MD} sensor was developed using HCF(III) and *Escherichia coli* by Pasco et al. in 2000 [72]. Since the study was reported, many BOD_{MD} biosensors have been intensively developed in recent years [73–76]. We have also developed several BOD_{MD} sensors. As the first, in our BOD_{MD} sensor which was reported in the same year as the report by Pasco et al., a batch system was employed using HCF (III) and *Pseudomonas fluorescens* biovar V which was isolated from a municipal sewage treatment plant [77] (see Table 3(VIa)). A combination chip consisting of a working electrode, on which *P. fluorescens* was immobilized, and a counter electrode was constructed. Under optimized conditions, synthetic sewage determined using the Organization for Economic Cooperation and Development (OECD) method was used as a standard solution approximating real wastewater to determine the BOD_{MD} using the sensor [78]. Using the OECD synthetic sewage (OSS), the influences of DO were investigated and the possibility of measuring BOD without the influences of DO was shown by this fact. By this work, the possibility of a new estimation method for BOD_{MD} determination was shown.

As the next step, a mobile type BOD sensor was developed [79] (see Table 3 (VIb)). In this study, storage conditions for the practical use of the BOD sensor chips were investigated and a handy type amperometer was fabricated for field monitoring. Under the conditions of limited nutrients, the microorganisms survive by metabolizing the endogenous substrates in the cell [80]. It has also been reported that bacteria in the "starved" condition can concurrently metabolize various exogenous substrates [81], which is advantageous for BOD measurements. In fact, it was found that the current response was increased by seven times compared to that before aeration at 131 mg O_2 L^{-1} of BOD_5 value and the linearity of the calibration curves was improved with up to 68 h of aeration. Finally, the sensor response decreased to approximately half the original after at least 35 days in storage.

As an application of these MD type BOD sensors, compost monitoring by BOD estimation was examined by improving the sensor system to a 10-channel system for rapid multidetermination of BOD [82] (see Table 3(Vic)). Screen-printed disposable sensor chips for single use were fabricated by incorporating *P. fluorescens* and HCF(III) immobilized in sodium alginate gel. In conclusion, the relative change in BOD sensor values determined using our system corresponded well with the BOD_5 values obtained using the standard BOD_5 method during 58 days of composting.

2.1.5 BOD$_{DM}$ Sensors

A DM system combining menadione and HCF(III) was studied using the eukaryote *Saccharomyces cerevisiae* [83, 84]. Baronian et al. revealed that menadione (vitamin K3; lipophilic mediator) can penetrate the outer cell membrane. Roustan et al. applied the DM system to measuring yeast activity in alcoholic fermentation [85]. Heiskanen et al. compared menadione bisulfite and menadione using yeast, and they revealed that hydrophobic menadione was superior to its water soluble bisulfite derivative for probing living cells [86]. Yeasts are easily handled, omnivorous to many kinds of organic substances, and stable even in saline solutions. Thus, we have applied the DM system to BOD estimation employing baker's yeast *S. cerevisiae* [87].

2.1.6 BOD$_{RC}$ Sensors

A simple, multiple simultaneous spectrophotometric method for new BOD determination using 2,6-dichlorophenolindophenol (DCIP) as the RCI was realized [88, 89] (see Table 3(VII)). The absorbance of DCIP decreases due to the metabolism of organic substances in aqueous samples by *P. fluorescens* (see Fig. 2). As a new technical tool, a microplate reader for a plate having 96 wells was used for multiple simultaneous spectrophotometric BOD$_{RC}$ estimation. Thus, blanks, standards, real samples, and replicates could be determined concurrently, ensuring the accuracy of this method. The process required only 20 min per determination, compared with BOD$_5$. This method gives a linear response ($r = 0.971$) to the OSS samples from 50 to 430 mg O_2 L^{-1}.

Next, a compact optical device and disposable sensing strips for the simultaneous determination of the BOD of multiple samples were developed [88]. Using this strip, real samples from a kitchen in our factory and a wastewater treatment plant were examined, and good correlations were observed between the BOD$_{RC}$ values derived using this system and those determined by the conventional BOD$_5$ method. In conclusion, this system is transportable and compact as well as being simple to operate and suitable for rapid, on-site measurements.

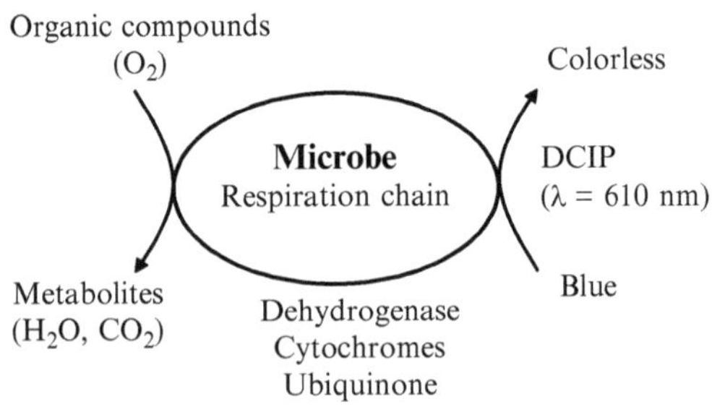

Fig. 2 Principle of BOD determination using DCIP by the spectrophotometric method

Recently, we have developed a highly sensitive and reproducible BOD_{RC} sensor using baker's yeast and a temperature-controlling system with a three-consecutive-stir unit [90]. A calibration curve for GGA concentration was obtained between 1.1 and 22 mg O_2 L^{-1} ($r = 0.988$, six points, $n = 3$) when the incubation mixture was incubated for only 10 min at 30°C. The reproducibility of the optical responses in the calibration curve was 1.77% (average of RSDs). This method was superior to the available BOD sensor (Central Kagaku Co.) in the detection limit (available BOD sensor's value, 2 mg O_2 L^{-1}), dynamic range (2–20 mg O_2 L^{-1}), reproducibility (5%), and measuring interval (30 min).

2.1.7 BOD_{SP} Sensors

According to Murakami et al., the SP technique was applied to a BOD sensor using *T. cutaneum* as a biosensing element [91] (see Table 3(VIII)). The biodegradation profiles of this biosensor were examined. The results in the BOD estimation of wastewaters obtained by both SP-based sensor and conventional BOD_5 methods corresponded well between 0 and about 180 mg O^2 L^{-1} BOD.

2.2 *Surfactant Sensors*

In synthetic anionic surfactants, linear alkylbenzene sulfonates (LASs) are most commonly used for the production of detergents. LAS is more easily biodegradable than nonlinear alkylbenzene sulfonate (ABS), and the biodegradation by microorganisms requires several days. However, LAS has toxicity in itself and also contributes to increases in the toxicity of other pollutants in the aquatic environment. A large amount of anionic surfactants such as LASs contained in domestic wastewater are allowed to flow into streams and rivers. In Japan, several ppm of anionic surfactants in a polluted river was reported. For the simple and rapid determination of LAS concentration, detergent biosensors indicating LAS were developed for river water monitoring [92, 93].

The first LAS biosensor was constructed using LAS degrading bacteria (strain A), which were isolated from activated sludge of a sewage treatment plant in Tokyo [92]. The LAS biosensor developed in this study is shown in Fig. 3. This is a reactor type sensor system consisting of immobilized LAS degrading bacteria and DO electrode. A sample solution was circulated in this flow type sensor system several times. The bacteria were immobilized in calcium alginate beads and the beads were packed into two columns (reactors). The two-column system was employed to raise the effects of LAS degradation. Then the LAS and DO concentrations in the sample were decreased by the LAS degrading bacteria packed in the two columns. By monitoring DO consumption in the sample, the LAS concentration was indirectly measured as well as showing the principle of the early BOD sensor. After optimization of the sensor system, a calibration curve was obtained between 0 and 4 mg L^{-1}

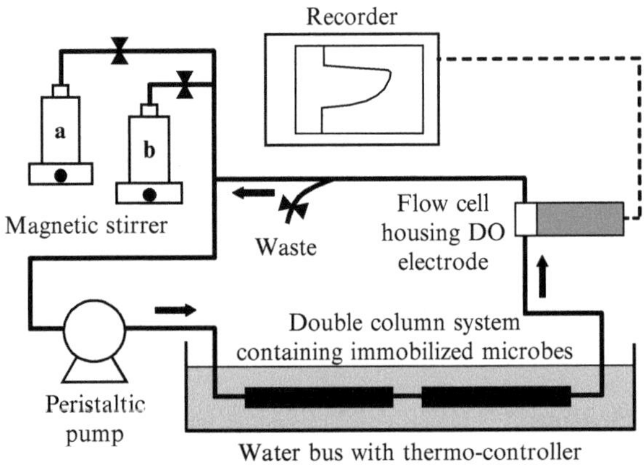

Fig. 3 Schematic diagram of the LAS measuring system using LAS degrading bacteria (strain A). a; Sample, b; buffer or $CaCl_2$ solution. Measurement was performed with in 15–30 min at 25–45°C with flow rate of 1.0 mL min^{-1}

LAS with a total of five points in the concentration range. Finally, the LAS degradability of strain A in this sensor system was confirmed by using the high-performance liquid chromatography (HPLC) method.

Subsequently, an LAS biosensor for river water monitoring was developed using the strain A and *T. cutaneum* [93]. In this study, to prevent the influence of coexisting substances dissolved in a real sample, a dual sensing system was constructed. Changes of LAS concentration in river water were monitored and mostly corresponded with the human life cycle in a day.

As other surfactant sensors, *Achromobacter* sp. and *Pseudomonas* sp. were used for determining anionic surfactants [94] and a fungal conidial cell (*Botrytis cinerea*) was used for determining cationic surfactants [95].

2.3 Chlorinated Hydrocarbon Sensors

Trichloroethylene (TCE) is an extensively used industrial solvent and, due to its broad application and high volatility, it has become a major contaminant of ground water. In Japan, it was found by the Environmental Agency in 1982 that many ground waters were contaminated by TCE with a higher concentration than the environmental standard allows (0.03 mg L^{-1}). To realize simple and rapid determination of TCE, whole cell biosensors have been developed using specific bacteria to degrade TCE [96, 97]. The bacteria *Pseudomonas aeruginosa* JI104 were isolated from soil near a gasworks [96]. This sensor system was optimized for sensitive TCE determination and obtained the linear concentration range for TCE from 0.1 to 4 mg L^{-1}. The response time was less than 10 min. This sensor characteristic makes it suitable for the detection of TCE in industrial wastewater. For improvement of this sensor system, a flow injection analysis (FIA) system was employed [97]. Finally, the sensor signals were successfully improved and linearly proportional to the concentration of TCE in the range from 0.03 to 2 mg L^{-1}. The sensor performance was checked on real samples, and this system showed good response in ground water, indicating its applicability for the on line monitoring at TCE contaminated areas.

As other chlorinated hydrocarbon sensors, a *Rhodococcus* sp. was used for determining chlorinated and brominated hydrocarbons [98] and transformed *E. coli* was used for determining halogenated organic acids [99].

2.4 Acid Rain Sensors

Acid rain is caused mainly by release into the atmosphere of oxides of sulfur (SO*x*) and nitrogen (NO*x*). For sulfite ion (SO_3^{2-}) sensing, *Thiobacillus thioparus* TK-m was used in a batch system [100] (see Table 4(Ia)). The SO_3^{2-} biosensor showed linear responses between 4 and 280 μM SO_3^{2-}. In this sensor system, it was found that sodium thiosulfate and sodium sulfate were interfering substances, although dimethyl sulfide did not influence the sensor response. To measure sulfate ion (SO_4^{2-}), a microbial sensor was also developed using *Thiobacillus ferrooxidans* strain 15 [101] (see Table 4(Ib)). The SO_4^{2-} biosensor showed linear responses between 4 and 200 μM SO_4^{2-}. In this sensor system, it was found that sodium nitrate was an interfering substance, although NaCl did not influence the sensor response. In addition, these two biosensors had poor stability.

To detect nitrogen dioxide (NO_2) gas, a high performance biosensor was developed using *Nitrobacter* sp. which was immobilized onto $CaCO_3$ particles [102,

Table 4 Characteristics of microbial cell sensors developed by our group for environmental monitoring

Target	Microbe	Indicator/ transducer	Measurement range (mg L^{-1})	Ref.
Acid rain				
(Ia) SO_3^{2-}	*T. thioparus* TK-m	DO/electrode	4–280 μM	[100]
(Ib) SO_4^{2-}	*T. ferrooxidans* strain 15	DO/electrode	4–200 μM	[101]
(Ic) O_2 (gaseous)	*Nitrobacter* sp.	DO/electrode	0.51–255	[102, 103]
(Id) $NaNO_2$	*Nitrobacter* sp.	DO/electrode	0.01–0.59 mM	[104]
Greenhouse effect gas				
(IIa) Methane (gaseous)	*M. flagellata* AJ 3670	DO/electrode	13.1 μM–6.6 mM	[105]
(IIb) Methane (gaseous)	*M. flagellata* AJ 3670	DO/electrode	3 μM–6.6 mM	[106]
(IIc) CO_2 (gaseous)	*Pseudomonas* sp. S-17	DO/electrode	5–200	[107]
(IId) CO_2 (gaseous)	*Pseudomonas* sp. S-17	DO/electrode	0.5–3.5 mM $NaHCO_3$	[108]
(IIe) CO_2 (gaseous)	Thermophilic bacteria	DO/electrode	1–8 mM	[109]
(IIf) CO_2 (gaseous)	Thermophilic bacteria	DO/electrode	3–12%	[109]

103] (see Table 4(Ic)). The NO_2 gas sensor showed linear responses between 0.51 and 255 mg L^{-1} NO_2 with no interfering substances. For liquid samples, a sodium nitrite ($NaNO_2$) biosensor was also developed using *Nitrobacter* sp. which was immobilized onto an acetylcellulose membrane [104] (see Table 4(Id)). The $NaNO_2$ biosensor showed linear responses between 10 and 590 μM $NaNO_2$ with no interfering substances.

2.5 The Greenhouse Effects Gas Sensors

Methane (CH_4) is known as a greenhouse effect gas and one of the indicators for the gasification process. *Methylomonas flagellata* AJ 3670 was used for CH_4 gas measurements and two gas-flow and reactor systems were constructed [105, 106] (see Table 4(IIa) and (IIb)). The first developed biosensor utilizing immobilized *M. flagellata* could measure 13.1 μM–6.6 mM CH_4 gas, and the next developed sensor using a cell suspension improved the linear range of CH_4 to 3 μM.

Carbon dioxide (CO_2) is known as a greenhouse effect gas and measurements of CO_2 are also important to medical diagnosis and the fermentation process. Using *Pseudomonas* sp. S-17 as a CO_2 utilizing atrophic bacterium, the first biosensor for CO_2 gas was studied for a liquid sample [107] (see Table 4(IIc)). After optimization, the biosensor showed linear responses between 5 and 200 mg L^{-1} CO_2. As the interfering substance, acetic acid was identified; however, formic acid, butanol, and ethanol did not affect the sensor response. Using the same microbe, a chip type biosensor was studied by developing a miniature O_2 electrode using microfabrication techniques for semiconductors [108] (see Table 4(IId)). This biosensor chip could measure from 0.5 to 3.5 mM $NaHCO_3$ within 3 min and use 10 times measurements. In addition, both types of CO_2 gas biosensors for liquid and gaseous samples were fabricated using thermophilic bacteria [109] (see Table 4(IIe) and (IIf)), respectively. Each biosensor measures between 1 and 8 mM CO_2 in liquid samples and between 3 and 12% CO_2 in gas samples.

2.6 Agricultural Agent Sensors

As a novel approach to detect atrazine, surface plasmon resonance (SPR) determination of P450 mRNA levels in *S. cerevisiae* was carried out [110]. This method was fast (15 min), highly sensitive, simple to use, and gave higher precision ($<2\%$) than the conventional enzyme-linked immunosorbent assay (ELISA) method. A linear relationship was obtained between the SPR response and atrazine concentration in the range of 1 ng^{-1} to 1 mg L^{-1} ($r = 0.993$). In addition, a highly sensitive toxicity measurement system was also developed utilizing quantification of induced P450 mRNAs by toxic chemicals and an FIA system based on an SPR method [111].

The DNA and peptide nucleic acid (PNA) probes containing a complementary sequence to a part of P450 mRNA hybridized to the probes were quantified. In this method, 10 ng L^{-1} (10 ppt) of atrazine was detected by using both DNA and PNA probes, and highly sensitive detection was achieved by amplifying the target P450 mRNA based on nucleic acid sequence-based amplification (NASBA).

As other agricultural agent sensors, organophosphorus aromatic nitro insecticides and *p*-nitrophenol were determined by the microbial-cell respiratory activity of *Acinetobacter calcoaceticus* A-122, *P. putida* C-11, and *P. putida* BA-11 [112], and organophosphate pesticides (nerve agents; parathion, paraoxon, and methyl parathion) were determined by using recombinant microorganism with surface expressed organophosphorus hydrolase [113]. Chloroform was detected by anaerobic microbial consortia [114], and polycyclic aromatic hydrocarbons were detected by *P. fluorescens* HK44 [115].

2.7 Toxicity Sensors

Toxicity assays are required for the prevention of environmental destruction and obstacles to the health of the human body. The toxicities are categorized into general toxicity and specific toxicity. The general toxicity includes the acute toxicity, i.e., 50% lethal dose (LD_{50}), subacute toxicity, and chronic toxicity. Toxicity assays have been widely required not only in the environmental field, but also in the food and diagnosis fields. Toxic substances are generally detected by using chemical analytical or physical systems, for example, ion-selective electrodes (ISEs) and HPLC. However, the toxicity of such chemical substances cannot be measured. Only biological methods using living biomaterials enable measurement of the toxicity. As the most convenient way to determine or detect the toxicity, organisms such as fish, birds, and microorganisms have been used, especially for detecting acute toxicity. In particular, several microorganisms were used for biosensing of toxicants such as cyanides, antibiotics, pesticides, and other acute toxicants.

To detect the chemical toxicity which affects organisms by their acute toxicities, several biosensors were developed. In this field, the development of toxicity sensors can be classified into three types, i.e., indicating metabolic activity, respiration activity, or biodegradability of toxic chemicals. The former kind of biosensor was developed for total toxicity sensing. The latter two kinds of biosensors were mainly developed for cyanide detection, and these biosensors have been reviewed in the literature [116].

For the assessment of toxicity in aquatic samples, in metabolic activity indicating methods, the "Microtox" method using luminous bacteria was developed and marketed as a bioassay system [117, 118]. However, the Microtox system is not suitable for online monitoring or continuous measurement of toxic compounds. Thus, in our group, several toxicity sensors were also developed using in vivo luminescence occurring in bacterial cells for the online monitoring or continuous measurement of toxic compounds.

Table 5 Characteristics of toxicant sensors

Target	Microbe	Indicator/ transducer	Measurement range (mg L^{-1})	Ref.
(Ia) Pesticide[a] and antibiotics[b]	*Recombinant E. coli* HB101 (pRSV)	PL/PM	0.05–10 (nonlinear)	[119]
(Ib) Antibiotics[c] and herbicides[d]	*Recombinant E. coli* 207.T4	PL/PM	0.5–10[c], 0.1–100[d]	[121]
(Ic) BC[e], SDS[f], Cr (VI)	*P. phosphoreum* MT10204	PL/PM	28.2–110 nM[e]	[122]
(Id) Antibiotics[c]	*Recombinant E. coli*	PL/PM	1.0–10	[123]
(IIa) CN^-	*S. cerevisiae* IFO 0337	DO/electrode	0.008–4	[125]
(IIb) CN^-	*S. cerevisiae*	DO/electrode	0.004–0.4	[126]
(IIc) CN^-	*S. cerevisiae*	Two DO/ electrode	0–0.4	[127]
(IIIa) CN^-	*P. fluorescens* NCIMB 11764	DO/electrode	0.004–0.4	[128]
(IIIb) CN^-	*P. fluorescens* NCIMB 11764	DO/electrode	0.04–0.4	[129]
(IIIc) CN^-	*P. fluorescens* NCIMB 11764	DO/electrode	0.02–0.4	[130]

[a]Sodium azide, fluoroacetic acid
[b]Chloramphenicol (CP), neomycin sulfate
[c]CP
[d]Metoxuron (Met), isoproturon, ioxynil, and propanol
[e]Benzalkonium chloride
[f]Sodium dodecyl sulfate

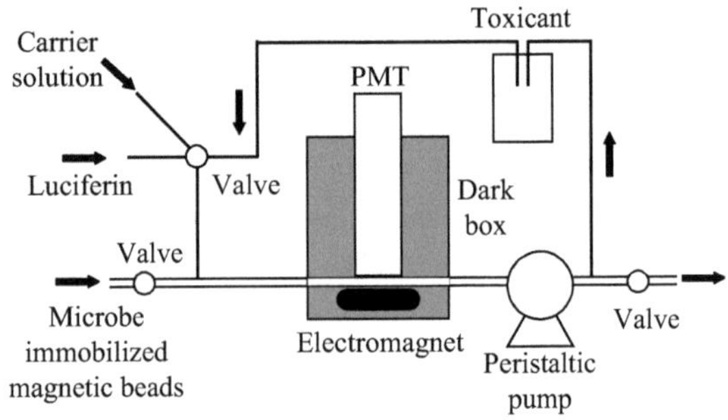

Fig. 4 Schematic diagram the batch-flow type toxicity detecting system

The first BL-based toxicity sensor was studied in 1991 [119, 120] (see Table 5 (Ia)). The toxicity was measured based on the decrease of in vivo luminescence intensity emitted by recombinant *E. coli* HB101 (pRSV), which was affected by cell metabolic inactivator. Toxicants such as fluoroacetic and sodium azide acid, which are components of adenosine triphosphate (ATP)-inhibiting pesticides, and antibiotics were detected at or below the μg L^{-1} level by this system. Using a similar measuring system and recombinant *E. coli* [121] (see Table 5(Ib)) or *P. phosphoreum* [122] (see Table 5(Ic)), antibiotics and pesticides were also detected [121, 122] (see Table 5(Ib) and (Ic)). Finally, the toxin sensor was improved to a batch-flow system using recombinant *E. coli* immobilized on magnetic beads [123] (see Table 5(Id)). The system was illustrated in Fig. 4. In this system, an electromagnet is placed below the photomultiplier (PM). The microbe-immobilized magnetic beads were incubated with toxicant solution for 30 min at 30°C. After washing with buffer solution, the beads flowed inside the tube and were held above the electromagnet. Chloramphenicol as antibiotic was detected in this system.

Cyanide is a deadly poison that inhibits respiration [124]. Nevertheless, it is widely utilized in industrial applications, especially for electroplating. Occasional accidents have taken place when industrial plants discharge cyanide into environmental water. The LD is in the range of 0.5–3.5 mg kg^{-1} body weight. Therefore, to regulate the discharge of cyanide into the environment, the Water Pollution Control Law in Japan stipulates 1 mg L^{-1} of cyanide as the maximum concentration of cyanide allowed in wastewater. Several cyanide biosensors have been developed following this principle [125–127] (see Table 5(II)).

2.7.1 Cyanide Sensor Indicating Respiration Activity of Aerobe

The first cyanide sensor was made by employing a DO electrode entrapped in yeast with an oxygen-permeable membrane, which was confirmed as a cyanide sensor [125]. The yeast, *S. cerevisiae* IFO 0337 was selected as a sensitive microorganism to cyanide. The yeast was entrapped between two porous cellulose nitrate membranes. The yeast in the microbial membrane can take up glucose, oxygen, and other nutritious substances through the porous membrane, being exposed to cyanide as well when it exists in the solution. This sensor was able to determine the cyanide ion (CN^-) with a linear range between 8.0 and 4,000 μg L^{-1}, which indicates that this cyanide biosensor employing a flow system and a reactor has possible applications for the monitoring of cyanide.

The second sensor employed a flow system [126]. Yeast was immobilized on glass beads acting as support; then the beads were packed into a column acting as a reactor. The system employed two electrodes, and the reactor was placed between them. This system was able to detect the CN^- with a linear range between 0 and 400 μg L^{-1}, under the conditions of a flow rate of 4.5 mL min^{-1} and 25°C. These results indicate the possible use of this sensor in the construction of a flow sensor system that can be applied to continuous monitoring for preventing the discharge of cyanide from wastewater.

The third sensor was developed for determining cyanide in river water using an improved previous sensor system [127]. The sensor employed a double electrode system and the reactor was set up between the two electrodes. Under optimized conditions, the sensor responded to CN^- at a linear range from 0 to 400 μg L^{-1}. As a result, the sensor was sensitive enough to detect cyanide contamination from industrial plants in river water.

Biosensors employing inhibition of microbial respiration are also affected by other toxic compounds, such as herbicides and pesticides. Therefore, this kind of sensor can be used for estimating total toxicity around a polluted water area. However, it is also important to improve the selectivity of the sensor as a cyanide detection system. Thus, subsequently, biosensors using cyanide degrading bacteria for the selective determination of cyanide were developed.

2.7.2 Cyanide Sensor Using Cyanide-Degrading Microbe

P. fluorescens NCIMB 11764 having cyanide oxidase and cyanase aerobically biodegrades cyanide as a sole nitrogen source. Cyanide oxidase produces cyanate from cyanide, consuming oxygen. Then the cyanate is hydrolyzed by cyanase, and carbon dioxide and ammonia are produced. Using this microbial degrading mechanism of cyanide, several batch or flow cell biosensors were developed [128–130] (see Table 5(III)).

The first microbial cell sensor using this bacterial degradation of cyanide was employed in a batch system [128]. Cyanide can be measured using a DO electrode to determine the decrease of DO by degrading cyanide of *P. fluorescens* NCIMB 11764. *P. fluorescens* NCIMB 11764 was immobilized between two membranes. The batch system of this sensor employs an electrode and a digital multimeter. After sensor characterization, the multiple effects of numerous substances on the sensor response were investigated using water from the Watarase River in Japan. The sensor responses to CN^- concentrations between 80 and 400 μg L^{-1} were determined. However, the membrane type sensor was affected by the concentrations of nutrients, such as GGA [128].

To improve selectivity, a cyanide-selective sensor using a gas-permeable membrane (PTFE; polytetrafluoroethylene) with a batch system was developed [129]. Cyanide in water forms molecular acid hydrogen cyanide (HCN) and/or free CN^-. Over pH 4, cyanides are easily converted into HCN. By using the gas-permeable membrane, *P. fluorescens* NCIMB 11764 immobilized in the microbial membrane was exposed only to volatilized gases such as HCN and oxygen. In this system, the nutrients did not affect the sensor response. A comparison of the JIS method (pyridine method) with the gas-phase biosensor was carried out using several samples of river waters, adding CN^- with a concentration between 40 and 400 μg L^{-1}. Consequently, a good correlation was obtained by the two methods (correlation coefficient 0.995, $n = 4$). In conclusion, the selectivity, sensitivity, and stability of the gas-phase biosensor were considered adequate for practical use, suggesting the usefulness of this sensor. Therefore, cyanide biosensors employing a reactor with a flow system have been implemented for the continuous monitoring of cyanide in river water.

A reactor and flow type cyanide sensor using an immobilized *P. fluorescens* NCIMB 11764 column was developed for environmental monitoring. *P. fluorescens* NCIMB 11764 (2.5 g) was entrapped in calcium alginate gel beads. Cyanide dissolved in the sample solution was degraded by *P. fluorescens* NCIMB 11764, and then consumed oxygen was detected by the DO electrode. After optimization, a calibration curve for CN^- was obtained with a linear response between 20 and 400 μg L^{-1} and a response time of 5 min. The result obtained had a wider range than that of the previous sensors. This sensor demonstrated its possible application for continuous detection of cyanide and online monitoring of river water, in the same manner as the flow and reactor type yeast sensor. The sensor response was obtained after at least 30 days. However, this sensor's selectivity needs to be improved. Therefore, the development of a gas-phase biosensor employing a flow system is required to avoid the influences of nutrients in river water.

2.7.3 Other Toxicity Sensors

As other developments, luminescence-based genetically engineered microbial cell sensors for water toxicity monitoring were developed using *S. cerevisiae* [131], freshwater bacterium [132], *E. coli* [133], or *E. coli* with an array system [134]. Luminescent bacteria were used for bioavailable toxic metals and metalloids from natural water samples [135]. Stress-responsive luminous bacteria were used for toxicants [136]. Green fluorescent protein-based genetically engineered microbial biosensors were developed using *S. cerevisiae* for genotoxicity monitoring [137]. For sediment quality control, application of microbial toxicity tests was reviewed [138]. For gas toxicity monitoring, *E. coli* GC1 [139] or GC2 [140] was employed. For heavy metal toxicity sensing, a cardiac cell-based biosensor was developed [141]. Early detection of wastewater toxicity was performed by using a microbial sensing system [142]. A respirometric biosensor system was developed using *Nitrosomonas europaea* for the detection of inhibitors of ammonia oxidation in wastewater [143]. As a new approach, biological toxicity detection was performed using an SPR system [144].

2.8 Other Environmental Sensors

Microbial cell sensors for other pollutants have also been developed. Phenolic compounds were detected by using *R. erythropolis* [145], *Moraxella* sp. [146], *P. putida* [147], and marine bioluminescent bacteria [148]. The photosynthetic activities of microalgae of *Chlorella* sp., *Scenedesmus* sp., and *Spirulina* sp. were measured by a system consisting of an O_2 electrode integrated with optical fibers [149]. Phosphate ion, which causes eutrophication, was detected using a luminescent cyanobacterial reporter strain [150]. Naphthalene was detected by *P. putida* BS238 carrying the naphthalene degradable plasmid pBS2 [151].

3 Food Quality Control

Microbial sensors for food quality control have been developed and the works were reviewed in several literature reports [152–158]. For fermentation process control in brewing or fuel production, monitoring of alcohol concentration is very important. For methanol (gaseous) monitoring in a liquid sample, *Trichosporon brassicae*

Table 6 Characteristics of microbial cell sensors developed by our group for food quality control

Target	Microbe	Indicator/transducer	Measurement range ($mg\ L^{-1}$)	Ref.
Alcohol				
(Ia) Methanol (gaseous)	*T. brassicae* CBS 6382	DO/electrode	2–22.5	[159]
(Ib) Methanol (gaseous)	Unidentified bacterium AJ 3993	DO/electrode	5.5–22.3	[160]
(Ic) Ethanol (gaseous)	*T. brassicae* CBS 6382	DO/electrode	2–22.3	[159–161]
(Id) Alcohol (gaseous)	*A. aceti* IAM 1802	pH/ISFET	0.1–70 mM	[163]
Organic acid				
(IIa) Formic acid (gaseous)	*C. butyricum* IFO 3847	H_2/fuel cell	10–1000	[168]
(IIb) Acetic acid (gaseous)	*T. brassicae* CBS 6382	DO/electrode	5–54	[159, 161, 169]
(IIc) Glutamic acid	*E. coli* ATCC 8739	CO_2/glass electrode	100–800	[171]
Ammonia				
(IIIa) Ammonia (gaseous)	Nitrifying bacteria	DO/electrode	3.5–42	[159, 161]
(IIIb) Ammonia (gaseous)	Nitrifying bacteria	DO/electrode	0.05–1.3	[180]
(IIIc) Ammonia (gaseous)	Nitrifying bacteria	DO/electrode	0.1–42	[181]
(IIId) Ammonia (gaseous)	Nitrifying bacteria	DO/electrode	0.45–10 mM	[182]
Vitamin				
(IVa) Vitamin B_1	*L. fermenti* ATCC 9338	NAD[a] or FAD[b]/ Fuel cell	0.001–0.05	[184]
(IVb) Vitamin B_{12}	*E. coli* 215	DO/electrode	0.005–0.025	[185, 186]
Sugar				
(Va) Glucose	*P. fluorescens* IFO 3081	DO/electrode	2.0–20	[189]
(Vb) Total assimilable sugars	*B. lactofermentum* AJ 1511	DO/electrode	<1 mM	[193]

[a]Nicotinamide adenine dinucleotide
[b]Flavin adenine dinucleotide

CBS 6382 [159] or unidentified bacterium AJ 3993 [160] was used respectively as sensing element in each biosensor system which employed an acetylcellulose membrane and the DO electrode (see Table 6(Ia) and (Ib)). Ethanol (gaseous) was also determined by *T. brassicae* CBS 6382 [159–161] (see Table 6(IIc)) or *Gluconobacter oxydans* with ferricyanide mediation [162]. *Acetobacter aceti* IAM 1802 was used for alcohol (gaseous) measurements with the ion-sensitive field-effect transistor (ISFET) element [163] (see Table 6(Id)). In addition, several microbial sensors were recently developed for ethanol determination using yeast [164–166]. For koji quality control in sake brewing as the fermentation process, *S. cerevisiae* K701 and K9 were employed in the SP device [167]. The pH change due to the production of organic acids in sake brewing was determined by the SP device. Yeast activity in alcoholic fermentation was measured by the mediator system combining HCF(III) and menadione [85].

For organic acid measurements, a specific microbial cell sensor for formic acid was developed [168] (see Table 6(IIa)). Formic acid is found in culture media, blood, urine, and gastric content as a product of many chemical reactions. It is a commonly occurring intermediate of cellular metabolism and is attracting attention as an intermediate of biomass conversion which is easily converted to hydrogen. For formic acid (gaseous) determination, the anaerobe *C. butyricum* IFO 3847 was immobilized in agar gel on an acetylcellulose membrane, and a fuel cell system for hydrogen detection was used. This system showed highly specific performance towards formic acid, although hydrogen in the sample influenced the formic acid measurements. An acetic acid biosensor is required in fermentation processes and was developed using *T. brassicae* [169] (see Table 6(IIb)). The biosensor enabled the determination of 5–54 mg L^{-1} acetic acid with long-term stability. For determination of tannic acid, the fungus *Aspergillus ustus* immobilized in poly (vinyl alcohol) was used in a batch system [170]. For amino acid determination, *E. coli* ATCC 8739 was used for L-glutamic acid [171] (see Table 6(IIc)), *Sarcina faecium* was used for L-arginine [172], *Bacterium cadavers* was used for L-aspartic acid [173], and *Sarcina flava* was used for L-glutamine [174]. The biosensor for L-glutamic acid determination utilized glutamate decarboxylase (GD) which was contained in *E. coli* cells and used a CO_2 gas sensor [171]. This sensor influenced oxygen and glutamine and did not influence other amino acids. Lactic acid was determined by a mediator type amperometric biosensor based on carbon paste electrodes modified with *S. cerevisiae* [175].

Simultaneous determination of sucrose, glucose, and lactose was performed by *S. cerevisiae*, *G. oxydans*, or *Kluyveromyces marxianus* [176]. Simultaneous determination of glucose and ethanol was performed by a nonselective microbial sensor for both ethanol and glucose using *G. oxydans* and a glucose electrode with glucose oxidase [177] and by *G. oxydans* or *Pichia methanolica*. The bacterial cells of *G. oxydans* were sensitive to both substrates, while the yeast cells of *P. methanolica* oxidized only ethanol [178]. For the simultaneous determination of mono- and disaccharides, a microbial cell sensor array with transport mutants of *E. coli* K12 was developed [179].

Ammonium ion or ammonia gas monitoring is not only required for food quality control, but also for environmental monitoring and medical diagnosis. Thus, several types of microbial cell sensors for them were studied. For the detection of ammonia (gaseous) in a liquid sample, nitrifying bacteria (unidentified) were used [159, 161, 180–182] (see Table 6(III)). The *Nitrobacter* sp. was immobilized on a gas-permeable Teflon membrane in a batch system and the biosensor measured 3.5–42 mg L^{-1} NH_3 [159]. Other microbial immobilization methods and measuring systems were also studied for the determination of gaseous targets [180–182]. *T. cutaneum* was immobilized in a membrane for long-term stability and was used for continuous monitoring of ammonium ion in sewage [183].

Several biosensors for determination of vitamins were also studied [184–188] (see Table 6(IV)). Vitamin B_1 (thiamine) in culture broth was measured by using

S. cerevisiae with a DO meter [188] or *Lactobacillus fermenti* ATCC 9338 with a fuel cell system [184]. In this study, a possible mechanism of current generation is discussed. A vitamin B_{12} determination system using a DO electrode and *E. coli* 215 was constructed [185, 186].

For sugar measurements, glucose was determined by *P. fluorescens* IFO 3081 [189] (see Table 6(Va)) and *Aspergillus niger* with mediator [190]. For xylose detection, *G. oxydans* cells were used in an amperometric device [191] and in the field-effect transistor (FET) device [192] respectively. For glucose sensing, *P. fluorescens* was immobilized in a collagen gel on a gas-permeable membrane [189]. Determination of total assimilable sugars (TASs) is required for online measurements of substrate in culture broths in the fermentation industry. TASs (sucrose, fructose, and glucose) were determined using *Brevibacterium lactofermentum* AJ 1511 which was immobilized in a dialysis membrane and a Teflon membrane [193] (see Table 6(Vb)).

4 Medical Diagnosis

Several types of microbial cell sensors were studied for medical diagnosis application. Hybrid biosensor systems using both materials of microbe and enzyme as sensing elements were mainly developed for clinical applications [194, 195]. Phenylalanine as an indicator of phenylketonuria was determined by incubation with *Leuconostoc mesenteroides* ATCC 8042 (6 h) followed by use of a lactate electrode (immobilized lactate oxidase (LOD) and an O_2 electrode) to measure the lactate produced [196] (see Table 7(Ia)). A linear relationship is obtained between the current decrease and the phenylalanine concentration between 75 and 600 $\mu g\ L^{-1}$. Next, to improve measuring time, the microbes were immobilized to agar on an acetylcellulose membrane and packed into a column as a reactor [197] (see Table 7 (Ib)). As a result, the measuring time was shortened to 90 min. A hybrid urea biosensor was developed using nitrifying bacteria from activated sludge and urease [198] (see Table 7(Ic)). This sensor measured ammonia (NH_3), which was produced from urea by urease reaction, by an electrode, immobilized nitrifying bacteria, and made possible the determination of 2–200 mM within 7 min. The determination of creatinine in serum and urine is a diagnostically important test for external dialysis. Then a hybrid creatinine sensor was developed using creatininase and nitrifying bacteria, and showed linear responses from 5 to 100 mg L^{-1} creatinine [199] (see Table 7(Id)). A hybrid biosensor for L-tyrosine determination, consisting of immobilized L-tyrosine decarboxylase and CO_2 utilizing chemoautotrophic bacterium, was developed and performed 10–50 μM detection for L-tyrosine [200] (see Table 7(Ie)).

Mutagen sensing is also important for medical diagnosis, and therefore several microbial cell sensors were studied for mutagen detection. Mutagen for the marker of cancer was detected using two types of *B. subtilis* [201] (see Table 7(IIa)). In a similar way, mutagens were detected by different immobilization of microbes to O_2 electrodes [202, 203] (see Table 7(IIb)). Other biosensors for mutagen detection

Table 7 Characteristics of microbial cell sensors developed by our group for medical diagnosis

Target	Microbe	Indicator/transducer	Measurement range ($mg\ L^{-1}$)	Ref.
Hybrid biosensors for constituent in body fluid				
(Ia) Phenylalanine	*L. mesenteroides* ATCC *8042*	DO (lactate)/electrode immob. LOD	0.075–0.6	[196]
(Ib) Phenylalanine	*L. mesenteroides* ATCC 8042	DO (lactate)/electrode immob. LOD	0.1–50	[197]
(Ic) Urea	Nitrifying bacteria (and urease)	DO (NH_4^+)/electrode	2–200 mM	[198]
(Id) Creatinine	Nitrifying bacteria (and creatininase)	DO (NH_4^+)/electrode	5–100 mg dL^{-1}	[199]
(Ie) L-Tyrosine	Chemoautotrophic bacterium	DO (CO_2)/electrode	10–50 mM	[200]
Mutagens				
(IIa) Mutagens	*B. subtilis M45 Rec*$^-$ and *B. subtilis H17 Rec*$^+$	DO/electrode	1.6 AF-2[a]	[201]
(IIb) Mutagens	*B. subtilis M45 Rec*$^-$ *and B. subtilis H17 Rec*$^+$	DO/electrode	1.6–2.8 AF-2	[202, 203]
(IIc) Mutagens	*S. typhimurium TA 100*	DO/electrode	0.001–0.006 AF-2	[203, 204]
(IId) Mutagens	*E. coli GY5027* (lysogenic strain*), GY5026*	DO/electrode	0.01–0.2 AF-2	[205]
(IIe) Mutagens	*E. coli GY5026* (lysogenic strain*)*	BL/PM	102–104 ng	[206]
Antibiotics				
(IIIa) Nystatin (antibiotics)	*S. cerevisiae*	DO/electrode	0.5–80 unit mL^{-1}	[207]
(IIIb) Cephalosporins (antibiotics)	*C. freundii B-0652*	H_2/glass electrode	62.5–300	[209]
Mirobial populations				
(IV) Mirobial populations	Five kinds of microbes were used	DO/electrode	$>10^6$ [b], $>10^5$ [c]	[216–218]

[a]2-(2-Furyl)-3-(5-nitro-2-furyl)acrylamide
[b]Bacteria cells
[c]Yeast cells

were also developed using *Salmonella typhimurium* TA 100 [204] (see Table 7(IIc)) and a combination of *E. coli* GY5027 (lysogenic strain) and *E. coli* GY5026 [205] (see Table 7(IId)). The latter type of biosensor was improved to a BL detection system using luminescent bacteria [206] (see Table 7(IIe)).

A nystatin biosensor was developed as an antibiotic sensor using *S. cerevisiae* and had a linear range between 0.5 and 80 unit mL^{-1} [207] (see Table 7(IIIa)). As another application, *E. coli* DPD2749 for mitomycin C was studied [208]. Then a collagen membrane was used instead of nylon net. As antibiotics, a cephalosporin measuring system was also developed using the hydrogen producing anaerobe of *Citrobacter freundii* B-0652 and a glass electrode (H_2) in a flow and reactor system [209] (see Table 7(IIIb)). The biosensor enabled measurement of cephalosporins between 62.5 and 300 mg L^{-1} 7-phenylacetylamidodesacetoxysporanic acid (phenylacetyl-7 ADCA), cephaloridine, cephalothin, and between 62.5 and 125 mg L^{-1} cephalosporin c, respectively. For diagnosis of leukemia, an enzyme activity measuring system in single cells has been developed using a microcell with a positionable dual electrode. Peroxidase activities in single neutrophils and single acute promyelocytic leukemia cells were measured by this method [210].

To prevent contamination, especially that caused by pathogenic bacteria in diagnostic and environmental fields, food processing and microbial detection techniques are required. Thus, several biosensing techniques have been developed [211, 212]. Ertl and Mikkelsen developed an electrochemical biosensing array employing lectinlipopolysaccharide recognition for the identification of microorganisms, and six microbial species including Gram-positive and Gram-negative microbes were examined for identification using this method [213]. Zhou et al. developed an enhanced fluorescent fiber-optic biosensor system for *S. typhimurium* detection [214]. As a microbial pathogen sensor, bacterial elicitor flagellin was detected with a bioelectronic portable system employing plant living cells [215]. Determination of microbial populations is also required in many fields; thus, several biosensors have been developed [216–219] (see Table 7(IV)).

5 Outlook

In this review of microbial cell sensors, many kinds of developments have been introduced. In most microbial cell sensors, the selectivity is a shortcoming. Thus, as molecular recognition elements, mainly genetically transformed microbes, have recently been engineered for selective measurements [3]. For example, BOD sensing yeast, *T. cutaneum* IFO 10466, was genetically transformed using a plasmid, pAN 7-1, for luminous BOD sensing [220]. *Pseudomonas stutzeri* AK61 containing the enzyme cyanidase was isolated from wastewater at a metal-plating plant by Watanabe et al. [221]. For cloning and expression, a gene encoding cyanidase was taken into *E. coli* [222]. Further, the cyanidase was improved to increase the *Km* value for cyanide by site-directed mutagenesis [223]. Magnetic bacteria may be applied for the development of a new type of microbial sensor utilizing magnetic particles (bacterial magnetic particles; BMPs) [224], because the BMPs have been widely applied to the development of biochips and other biosensing devices [225]. By progression of this microbial cell sensor technique, the application will be expanded, especially for environmental monitoring and quality control fields as described before.

References

1. Karube I, Mitsuda S, Matsunaga T, Suzuki S (1977) J Ferment Technol 55:243
2. Nakamura H, Shimomura-Shimizu M, Karube I (2008) Adv Biochem Eng/Biotechnol 109:351
3. Rogers KR (2006) Anal Chim Acta 568:222
4. Karube I, Tamiya E (1987) Pure Appl Chem 59:545
5. Karube I, Nomura Y, Arikawa Y (1995) Trends Anal Chem 14:295
6. Yano K, Sasaki S, Nomura Y, Ikebukuro K, Karube I (1998) Enzyme Eng 14:23
7. Belkin S (2003) Curr Opin Microbiol 6:206
8. Karube I, Matsunaga T, Suzuki S (1977) J Solid-Phase Biochem 2:97
9. Karube I, Matsunaga T, Mitsuda S, Suzuki S (1977) Biotechnol Bioeng 19:1535
10. Suzuki S, Karube I (1978) Enzyme Eng 4:329
11. Hikuma M, Suzuki H, Yasuda T, Karube I, Suzuki S (1979) Eur J Appl Microbiol Biotechnol 8:289
12. Bond RG, Straub CP (eds) (1973) Handbook of environmental control. CRC, Cleveland
13. Japanese Industrial Standard Committee Testing Methods for Industrial Waste Water, edited JIS, Tokyo (1993), JIS K 3602
14. Central Kagaku Co. http://www.aqua-ckc.jp/ (Sep. 2009)
15. Yang Z, Suzuki H, Sasaki S, Karube I (1996) Appl Microbiol Biotechnol 46:10
16. Yang Z, Suzuki H, Sasaki S, McNiven S, Karube I (1997) Sens Actuators B 45:217
17. Yang Z, Sasaki S, Karube I, Suzuki H (1997) Anal Chim Acta 357:41
18. Chee GJ, Nomura Y, Ikebukuro K, Karube I (2000) Biosens Bioelectron 15:371
19. Chee GJ, Nomura Y, Karube I (1999) Anal Chim Acta 379:185
20. Chee GJ, Nomura Y, Karube I (1999) Anal Chim Acta 394:65
21. Chee GJ, Nomura Y, Ikebukuro K, Karube I (2001) Sens Actuators B 80:15
22. Chee GJ, Nomura Y, Ikebukuro K, Karube I (2005) Biosens Bioelectron 21:67
23. Chee GJ, Nomura Y, Ikebukuro K, Karube I (2007) Biosens Bioelectron 22:3092
24. Kang KH, Jang JK, Pham TH, Moon H, Chang IS, Kim BH (2003) Biotechnol Lett 25:1357
25. Murakami K, Hasegawa K, Watanabe H, Komori K (1978) Public Works Research Institute Technical Memorandum 1421
26. Tanaka H, Nakamura E, Minamiyama Y, Toyoda T (1994) Water Sci Technol 30:215
27. Lehmann M, Chan C, Lo A, Lung M, Tag K, Kunze G, Riedel K, Gruendig B, Renneberg R (1999) Biosens Bioelectron 14:295
28. Chan C, Lehmann M, Tag K, Lung M, Gotthard K, Riedel K, Gruendig B, Renneberg R (1999) Biosens Bioelectron 14:131
29. Chan C, Lehmann M, Chan K, Chan P, Chan C, Gruendig B, Kunze G, Renneberg R (2000) Biosens Bioelectron 15:343
30. Riedel K, Renneberg R, Kuhn M, Scheller F (1988) Appl Microbiol Biotechnol 28:316
31. Ohki A, Shinohara K, Ito O, Naka K, Maeda S, Sato T, Akano H, Kato N, Kawamura Y (1994) Int J Environ Anal Chem 56:261
32. Kulys J, Kadziauskiene K (1980) Biotechnol Bioeng 22:221
33. Kim MN, Park KH (2001) Sens Actuators B 80:9
34. Kim MN, Kwon HS (1999) Biosens Bioelectron 14:1
35. Sangeetha S, Sugandhi G, Murugesan M, Madhav VM, Berchmans S, Rajasekar R, Jeyakumar D, Rao GP (1996) Electroanalysis 8:698
36. Tan TC, Neoh KG, Li F (1992) Sens Actuators B 8:167
37. Li F, Tan TC, Lee YK (1994) Biosens Bioelectron 9:197
38. Li F, Tan TC (1994) Biosens Bioelectron 9:315
39. Heim S, Schnieder I, Binz D, Vogel A, Bilitewski U (1999) Biosens Bioelectron 14:187
40. Jia J, Tang M, Chen X, Qi L, Dong S (2003) Biosens Bioelectron 18:1023
41. Sakai Y, Abe N, Takeuchi S, Takahashi F (1995) J Ferment Bioeng 80:300

42. Liu J, Bjornsson L, Mattiasson B (2000) Biosens Bioelectron 14:883
43. Stand SE, Carlson DA (1984) J Water Pollut Control Fed 56:464
44. Rastogi S, Kumar A, Mehra NK, Makhijani SD, Manoharan A, Gangal V, Kumar R (2003) Biosens Bioelectron 18:23
45. Rastogi S, Rathee P, Saxena TK, Mehra NK, Kumar R (2003) Curr Appl Phys 3:191
46. Kim HB, Chang IS, Cheol GG, Park HS, Kim HJ (2003) Biotechnol Lett 25:541
47. Chang IS, Jang JK, Gil GC, Kim M, Kim HJ, Cho BW, Kim BH (2004) Biosens Bioelectron 19:607
48. Suzuki M, Takahashi S, Ishibashi M, Natsume K (1995) Sens Mater 7:159
49. Suzuki M, Ogata K, Tanaka M, Natsume K (1995) Denki Kagaku 63:1128
50. Karube I, Yokoyama K, Sode K, Tamiya E (1989) Anal Lett 22:791
51. Tan TC, Qian Z (1997) Sens Actuators B 40:65
52. Qian Z, Tan TC (1998) Chem Eng Sci 53:3281
53. Qian Z, Tan TC (1998) Water Res 32:801
54. Qian Z, Tan TC (1998) Water Res 33:2923
55. Tan TC, Wu C (1999) Sens Actuators B 54:252
56. Viaopoulou E, Melidis P, Kampragou E, Aivasidis A (2005) Biosens Bioelectron 21:365
57. Konig A, Bachmann TT, Metzger JW, Schmid RD (1999) Appl Microbiol Biotechnol 51:112
58. Gil GC, Chang IS, Kim BH, Kim M, Jang JK, Park HS, Kim HJ (2003) Biosens Bioelectron 18:327
59. Moon H, Chang IS, Kang KH, Jang JK, Kim BH (2004) Biotechnol Lett 26:1717
60. Chang IS, Moon H, Jang JK, Kim BH (2005) Biosens Bioelectron 20:1856
61. Kumlanghan A, Liu J, Thavarungkul P, Kanatharana P, Mattiasson B (2007) Biosens Bioelectron 22:2939
62. Hyun CK, Inoue N, Tamiya E, Takeuchi T, Karube I (1993) Biotechnol Bioeng 41:1107
63. Hastings JW, Potrikus CJ, Cupta SC, Kurfurst M, Makemson JC (1985) Adv Microb Physiol 26:235
64. Sakaguchi T, Morita Y, Yamasaki M, Iwanaga J, Beppu K, Maeda H, Morita Y, Tamiya E (2007) Biosens Bioelectron 22:1345
65. Bennetto HP, Stiring JL, Tanaka K, Vega CA (1983) Biotechnol Bioeng 25:559
66. Roller SD, Bennetto HP, Delaney GM, Mason JR, Stirling JL, Thurston CF (1984) J Chem Technol Biotechnol 34B:3
67. Nishikawa S, Sakai S, Karube I, Matsunaga T, Suzuki S (1982) Appl Environ Microbiol 43:814
68. Learoyd SA, Kroll RG, Thurston CF (1992) J Appl Biotechnol 72:479
69. Kalab T, Skladal P (1994) Electroanalysis 6:1004
70. Kubiak WW, Wang J (1989) Anal Chim Acta 221:43
71. Katrlik J, Brandsteter R, Svorc J, Rosenberg M, Miertus S (1997) Anal Chim Acta 356:217
72. Pasco N, Baronian K, Jeffries C, Hay J (2000) Appl Microbiol Biotechnol 53:613
73. Morris K, Catterall K, Zhao H, Pasco N, John R (2001) Anal Chim Acta 442:129
74. Catterall K, Morris K, Gladman C, Zhao H, Pasco N, Richard John R (2001) Talanta 55:1187
75. Pasco N, Hay J, Webber J (2001) Biomarkers 6:83
76. Catterall K, Zhao H, Pasco N, John R (2003) Anal Chem 75:2584
77. Yoshida N, Yano K, Morita T, McNiven SJ, Nakamura H, Karube I (2000) Analyst 12:2280
78. OECD Guideline for Testing of Chemicals (1993) 209:1
79. Yoshida N, Hoashi J, Morita T, McNiven SJ, Nakamura H, Karube I (2001) J Biotechnol 88:269
80. Cabral JPS (1994) Microbios 78:47
81. Leefeldt RH, Martin A (1980) J Bacteriol 142:645
82. Yoshida N, Hoashi J, Morita T, McNiven SJ, Yano K, Yoshida A, Nakamura H, Karube I (2001) Analyst 126:1751
83. Yashiki Y, Yamashoji S (1996) J Ferment Bioeng 82:319
84. Baronian KHR, Downard AJ, Lowen PK, Pasco N (2002) Appl Microbiol Biotechnol 60:108

85. Roustan JL, Sablayrolles JM (2003) J Biosci Bioeng 96:434
86. Heiskanen A, Yakovleva J, Spegel C, Taboryski R, Koudelka-Hep M, Emneus J, Ruzgas T (2004) Electrochem Commun 6:219
87. Nakamura H, Suzuki K, Ishikuro H, Kinoshita S, Koizumi R, Okuma S, Gotoh M, Karube I (2007) Talanta 72:210
88. Yoshida N, McNiven SJ, Morita T, Nakamura H, Karube I (2002) Anal Lett 35:1541
89. Yoshida N, McNiven SJ, Yoshida A, Morita T, Nakamura H, Karube I (2001) Field Anal Chem Technol 5:222
90. Nakamura H, Kobayashi S, Hirata Y, Suzuki K, Mogi Y, Karube I (2007) Anal Biochem 369:168
91. Murakami Y, Kikuchi T, Yamaura A, Sakaguchi T, Yokoyama K, Tamiya E (1998) Sens Actuators B 53:163
92. Nomura Y, Ikebukuro K, Yokoyama K, Takeuchi T, Arikawa Y, Ohno S, Karube I (1994) Anal Lett 27:3095
93. Nomura Y, Ikebukuro K, Yokoyama K, Takeuchi T, Arikawa Y, Ohno S, Karube I (1998) Biosens Bioelectron 13:1047
94. Taranova L, Semenchuk I, Manolov T, Iliasov P, Reshetilov A (2002) Biosens Bioelectron 17:635
95. Cabral SMJCS, Vasconcelos MTSD, Cabral JPS (1995) J Microbiol Methods 24:11
96. Han TS, Kim YC, Sasaki S, Yano K, Ikebukuro K, Kitayama A, Nagamune T, Karube I (2001) Anal Chim Acta 431:225
97. Han TS, Sasaki S, Yano K, Ikebukuro K, Kitayama A, Nagamune T, Karube I (2002) Talanta 57:271
98. Peter J, Hutter W, Stollnberger W, Hampel W (1996) Biosens Bioelectron 11:1215
99. Taube M, Rosen R, Belkin S (2001) Talanta 55:959
100. Marty JL, Sode K, Karube I (1992) Electroanalysis 4:249
101. Sasaki S, Yokoyama K, Tamiya E, Karube I, Hayashi C, Arikawa Y, Numata M (1997) Anal Chim Acta 347:275
102. Okada T, Karube I, Suzuki S (1983) Biotechnol Bioeng 25:1641
103. Suzuki S, Karube I (1982) Enzyme Eng 6:387
104. Karube I, Okada T, Suzuki S, Suzuki H, Hikuma M, Yasuda Y (1982) Eur J Appl Microbiol Biotechnol 15:127
105. Okada T, Karube I, Suzuki S (1981) Eur J Appl Microbiol Biotechnol 12:102
106. Karube I, Okada T, Suzuki S (1982) Anal Chim Acta 135:61
107. Suzuki H, Tamiya E, Karube I (1987) Anal Chim Acta 199:85
108. Suzuki H, Kojima N, Sugama A, Takei F, Ikegami K, Tamiya E, Karube I (1989) Electroanalysis 1:305
109. Suzuki H, Tamiya E, Karube I, Oshima T (1988) Anal Lett 21:1323
110. Lym TK, Oyama M, Ikebukuro K, Karube I (2000) Anal Chem 72:2856
111. Oyama M, Ikeda T, Lim TK, Ikebukuro K, Masuda Y, Karube I (2001) Biotechnol Bioeng 71:217
112. Guliy OI, Ignatov OV, Makarov OE, Ignatov VV (2003) Biosens Bioelectron 18:1005
113. Mulchandani P, Chen W, Mulchandani A, Wang J, Chen L (2001) Biosens Bioelectron 16:433
114. Pollice A, Rozzi A, Concetta Tomei M, Di Pinto AC, Laera G (2001) Water Res 35:1179
115. Ripp S, Nivens DE, Werner C, Sayler GS (2000) Appl Microbiol Biotechnol 53:736
116. Arikawa Y, Ikebukuro K, Karube I (1998) Microbial biosensors based on respiratory inhibition. In: Mulchandani A, Rogers KR (eds) Methods in biotechnology, vol 6. Humana, New Jersey
117. Bulich AA, Isenberg DL (1981) ISA Trans 20:29
118. Kamlet MJ, Doherty RM, Veith GD, Taft RW (1986) Environ Sci Technol 20:690
119. Lee SM, Suzuki M, Tamiya E, Karube I (1991) Anal Chim Acta 244:201
120. Lee SM, Suzuki M, Tamiya E, Karube I (1991) Chem Express 6:415

121. Lee SM, Suzuki M, Tamiya E, Karube I (1992) Anal Chim Acta 257:183
122. Lee SM, Sode K, Nakanishi K, Marty JL, Tamiya E, Karube I (1992) Biosens Bioelectron 7:273
123. Karube I, Asakura J, Arikawa Y, Takei T, Sasaki S, Suzuki M, Tamiya E (1995) Denki Kagaku 63:1134
124. Ikebukuro K, Nakamura H, Karube I (1999) Cyanides. In: Nollet LML (ed) Handbook of water analysis. Marcel Dekker, New York
125. Nakanishi K, Ikebukuro K, Karube I (1996) Appl Biochem Biotechnol 60:97
126. Ikebukuro K, Honda M, Nakanishi K, Nomura Y, Masuda Y, Yokoyama K, Yamauchi Y, Karube I (1996) Electroanalysis 8:876
127. Ikebukuro K, Miyata A, Cho SJ, Nomura Y, Chang SM, Yamauchi Y, Hasebe Y, Uchiyama S, Karube I (1996) J Biotechnol 48:73
128. Lee JI, Karube I (1995) Anal Chim Acta 313:69
129. Lee JI, Karube I (1996) Biosens Bioelectron 11:1147
130. Lee JI, Karube I (1996) Electroanalysis 8:1117
131. Hollis RP, Killham K, Glover LA (2000) Appl Environ Microbiol 66:1676
132. Cho JC, Park KJ, Ihm HS, Park JE, Kim SY, Kang I, Lee KH, Jahng D, Lee DH, Kim SJ (2004) Biosens Bioelectron 20:338
133. Lee JH, Gu MB (2005) Biosens Bioelectron 20:1744
134. Lee JH, Mitchell RJ, Kim BC, Cullen DC, Gu MB (2005) Biosens Bioelectron 21:500
135. Tauriainen SM, Virta MPJ, Karp MT (2000) Water Res 34:2661
136. Belkin S, Smulski DR, Dadon S, Vollmer AC, Van Dyk TK, Larossa RA (1997) Water Res 31:3009
137. Billinton N, Baker MG, Michel CE, Knight AW, Heyer WD, Goddard NJ, Fielden PR, Walmsley RM (1998) Biosens Bioelectron 13:831
138. Beelen P (2003) Chemosphere 53:795
139. Gil GC, Mitchell RJ, Chang ST, Gu MB (2000) Biosens Bioelectron 15:23
140. Gil GC, Kim YJ, Gu MB (2002) Biosens Bioelectron 17:427
141. Liu Q, Cai H, Xu Y, Xiao L, Yang M, Wang P (2007) Biosens Bioelectron 22:3224
142. Liao JD, Wang SH, Hsu DJ (2001) Sens Actuators B 72:167
143. Cui R, Chung WJ, Jahng D (2005) Biosens Bioelectron 20:1788
144. Choi JW, Park KW, Lee DB, Lee W, Lee WH (2005) Biosens Bioelectron 20:2300
145. Emelyanova EV, Reshetilov AN (2002) Process Biochem 37:683
146. Mulchandani P, Lei Y, Chen W, Wang J, Mulchandani A (2002) Anal Chim Acta 470:79
147. Timur S, Pazarlioglu N, Pilloton R, Telefoncu A (2003) Talanta 61:87
148. Philp JC, Balmand S, Hajto E, Bailey MJ, Wiles S, Whiteley AS, Lilley AK, Hajto J, Dunbar SA (2003) Anal Chim Acta 487:61
149. Takeuchi T, Yokoyama K, Kobayashi K, Suzuki M, Tamiya E, Karube I, Utsunomiya K, Imai O, Masuda Y (1993) Anal Chim Acta 276:65
150. Schreiter PPY, Gillor O, Post A, Belkin S, Schmid RD, Bachmann TT (2001) Biosens Bioelectron 16:811
151. Reshetilov AN, Iliasov PV, Filonov AE, Gayazov RR, Kosheleva IA, Boronin AM (1997) Process Biochem 32:487
152. Karube I, Suzuki S (1983) Application of biosensor in fermentation processes. In: Tsao GT (ed) Annual reports on fermentation processes, vol 6. Academic, New York
153. Karube I (1985) Biosensors in fermentation and environmental control. In: Arbor BA (ed) Handbook of biotechnology. Science Publishing, New Jersey
154. Karube I, Tamiya E (1987) Food Biotechnol 1:147
155. Karube I, Sode K (1989) Biosensors for fermentation process control. In: Ghose TK (ed) Bioprocess engineering: the first generation. Chapman Hall, London
156. Karube I, Sode K (1991) Microbial sensors for process and environmental control. In: Wise DL (ed) Bioinstrumentation and biosensors. Marcel Dekker, New York
157. Ohashi E, Karube I (1993) Food Control 4:183

158. Karube I (1994) Application of biosensors in measurement of food. In: Karube I (ed) On-line sensors for food processing. Gordon and Breach, Yverdon
159. Karube I, Suzuki S, Okada T, Hikuma M (1980) Biochimie 62:567
160. Hikuma M, Kubo T, Yasuda T, Karube I (1979) Biotechnol Bioeng 21:1845
161. Karube I, Suzuki S (1980) Enzyme Eng 5:263
162. Tkac J, Vostiar I, Gemeiner P, Sturdik E (2002) Bioelectrochemistry 56:127
163. Kitagawa Y, Tamiya E, Karube I (1987) Anal Lett 20:81
164. Tkac J, Vostiar I, Gorton L, Gemeiner P, Sturdik E (2003) Biosens Bioelectron 18:1125
165. Rotariu L, Bala C, Magearu V (2004) Anal Chim Acta 513:119
166. Liao MH, Guo JC, Chen WC (2006) J Magn Magn Mater 304:e421
167. Chiyo T, Matsui K, Murakami Y, Yokoyama K, Tamiya E (2001) Biosens Bioelectron 16:1021
168. Matsunaga T, Karube I, Suzuki S (1980) Eur J Appl Microbiol Biotechnol 10:253
169. Hikuma M, Kubo T, Yasuda T, Karube I, Suzuki S (1979) Anal Chim Acta 109:33
170. Zhao YB, Wen ML, Liu SQ, Liu ZH, Zhang WD, Yao Y, Wang CY (1998) Microchem J 60:201
171. Hikuma M, Obana H, Yasuda T, Karube I, Suzuki S (1980) Anal Chim Acta 116:61
172. Rechnitz GA, Kobos RK, Riechel SJ, Gebauer R (1977) Anal Chim Acta 94:357
173. Kobos RK, Riechel SJ (1977) Anal Lett 10:751
174. Rechnitz GA, Riechel SJ, Kobos RK, Meyerhoff ME (1978) Science 199:440
175. Garjonytea R, Melvydasb V, Malinauskas A (2006) Bioelectrochemistry 68:191
176. Svitel J, Curilla O, Tkac J (1998) Biotechnol Appl Biochem 27:153
177. Reshetilov AN, Lobanov AV, Morozova NO, Gordon SH, Greene RV, Leathers TD (1998) Biosens Bioelectron 13:787
178. Lobanov AV, Borisov IA, Gordon SH, Greene RV, Leathers TD, Reshetilov AN (2001) Biosens Bioelectron 16:1001
179. Held M, Schuhmann W, Jahreis K, Schmidt HL (2002) Biosens Bioelectron 17:1089
180. Hikuma M, Suzuki H, Yasuda T, Karube I, Suzuki S (1980) Anal Chem 52:1020
181. Karube I, Okada T, Suzuki S (1981) Anal Chem 53:1852
182. Okada T, Karube I, Suzuki S (1982) Anal Chim Acta 135:159
183. Ikeda M, Hachiya H, Ito S, Asano Y, Imato T (1998) Biosens Bioelectron 13:531
184. Matsunaga T, Karube I, Suzuki S (1978) Anal Chim Acta 98:25
185. Karube I, Wang Y, Tamiya E, Kawarai M (1987) Anal Chim Acta 199:93
186. Wang E, Karube I, Suzuki S (1986) Chin J Biotechnol 2:51
187. Kalman A, Caelen I, Svorc J (2006) J AOAC Int 89:819
188. Akyilmaz E, Yasa I, Dinckaya E (2006) Anal Biochem 354:78
189. Karube I, Mitsuda S, Suzuki S (1979) Eur J Appl Microbiol Biotechnol 7:343
190. Katrlk J, Brandsteter R, Svorc J, Rosenberg M, Miertus S (1997) Anal Chem 356:217
191. Reshetilov AN, Iliasov PV, Donova MV, Dovbnya DV, Boronin AM, Leathers TD, Greene RV (1997) Biosens Bioelectron 12:241
192. Reshetilov AN, Donova MV, Dovbnya DV, Boronin AM, Leathers TD, Greene RV (1996) Biosens Bioelectron 11:401
193. Hikuma M, Obana H, Yasuda T, Karube I, Suzuki S (1980) Enzyme Microb Technol 2:234
194. Karube I, Kubo I, Suzuki S (1984) Ann NY Acad Sci 424:508
195. Karube I (1988) Methods Enzymol 137:131
196. Karube I, Matsunaga T, Teraoka N, Suzuki S (1980) Anal Chim Acta 119:271
197. Matsunaga T, Karube I, Teraoka N, Suzuki S (1981) Anal Chim Acta 127:245
198. Okada T, Karube I, Suzuki S (1982) Eur J Appl Microbiol Biotechnol 14:149
199. Kubo I, Karube I, Suzuki S (1983) Anal Chim Acta 151:371
200. Suzuki H, Tamiya E, Karube I (1989) Anal Lett 22:15
201. Karube I, Matsunaga T, Nakahara T, Suzuki S (1981) Anal Chem 53:1024
202. Karube I, Suzuki S (1982) Enzyme Eng 6:417
203. Karube I (1984) Trends Anal Chem 3:40

204. Karube I, Nakahara T, Matsunaga T, Suzuki S (1982) Anal Chem 54:1725
205. Karube I, Sode K, Suzuki M, Nakahara T (1989) Anal Chem 61:2388
206. Lee SM, Suzuki M, Kumagai M, Ikeda H, Tamiya E, Karube I (1992) Anal Chem 64:1755
207. Karube I, Matsunaga T, Suzuki S (1979) Anal Chim Acta 109:39
208. Gu MB, Gil GC, Kim JH (1999) Biosens Bioelectron 14:355
209. Matsumoto K, Seijo H, Watanabe T, Karube I, Satoh I, Suzuki S (1979) Anal Chim Acta 105:429
210. Gao N, Zhao M, Zhang X, Jin W (2006) Anal Chem 78:231
211. Hobsen NS, Tothill IE, Turner APF (1996) Biosens Bioelectron 11:455
212. Ivinski D, Abdell-Hamid I, Atanasov P, Wilkins E (1999) Biosens Bioelectron 14:599
213. Ertl P, Mikkelsen SR (2001) Anal Chem 73:4241
214. Zhou C, Pivarnik P, Rand G, Lecher SV (1998) Biosens Bioelectron 13:495
215. Oczkowski T, Zwierkowska E, Bartkowiak S (2007) Bioelectrochemistry 70:192
216. Matsunaga T, Karube I, Nakahara T, Suzuki S (1981) Eur J Appl Microbiol Biotechnol 12:97
217. Ishimori Y, Karube I, Suzuki S (1981) Appl Environ Microbiol 42:632
218. Matsunaga T, Karube I, Suzuki S (1979) Appl Environ Microbiol 37:117
219. Ramsayl G, Turner APF (1988) Anal Chim Acta 215:61
220. Sakaguchi T, Amari S, Nagashio N, Murakami Y, Yokoyama K, Tamiya E (1998) Biotechnol Lett 20:851
221. Watanabe A, Yano K, Ikebukuro K, Karube I (1998) Microbiology 144:1677
222. Watanabe A, Yano K, Ikebukuro K, Karube I (1998) Appl Microb Biotechnol 50:93
223. Watanabe A, Yano K, Ikebukuro K, Karube I (1998) Biochim Biophys Acta 1382:1
224. Sakaguchi T, Burgess JG, Matsunaga T (1993) Nature 365:47
225. Matsunaga T, Sakaguti T (2000) J Biosci Bioeng 90:1

Adv Biochem Engin/Biotechnol (2010) 118: 31–64
DOI: 10.1007/10_2009_9

Published online: 19 June 2009

Whole-Cell Bioreporters for the Detection of Bioavailable Metals

Anu Hynninen and Marko Virta

Abstract Whole-cell bioreporters are living microorganisms that produce a specific, quantifiable output in response to target chemicals. Typically, whole-cell bioreporters combine a sensor element for the substance of interest and a reporter element coding for an easily detectable protein. The sensor element is responsible for recognizing the presence of an analyte. In the case of metal bioreporters, the sensor element consists of a DNA promoter region for a metal-binding transcription factor fused to a promoterless reporter gene that encodes a signal-producing protein.

In this review, we provide an overview of specific whole-cell bioreporters for heavy metals. Because the sensing of metals by bioreporter microorganisms is usually based on heavy metal resistance/homeostasis mechanisms, the basis of these mechanisms will also be discussed. The goal here is not to present a comprehensive summary of individual metal-specific bioreporters that have been constructed, but rather to express views on the theory and applications of metal-specific bioreporters and identify some directions for future research and development.

Keywords Reporter gene • Bioavailability • Microbial • Metal resistance • Luciferase • *lux* • *luc* • Microbial ecology

A. Hynninen and M. Virta
Department of Applied Chemistry and Microbiology, University of Helsinki,
P.O.Box 56, 00014 Helsinki, Finland
e-mail: marko.virta@helsinki.fi

Contents

1 Introduction

Whole-cell bioreporters are living microorganisms that produce a specific quantifiable output in response to target chemicals. On the basis of detection specificity, bioreporters can be broadly divided into three groups: nonspecific, semispecific, and specific. Nonspecific bioreporters are useful for measuring the toxicity of samples but do not provide any information about the identity of contaminants. Recombinant semispecific bioreporters allow the detection of compounds that cause certain cellular responses, such as stress [1–3] or DNA damage [4, 5] without further characterizing the compound. Unlike nonspecific bioreporters, semispecific bioreporters contain one or more sensor elements, which usually consist of a promoter that is activated in response to global changes in the cell. Specific bioreporters, first described in [6] respond only to a certain compound or class of compounds, and therefore permit a quantitative analysis of contaminant concentration.

In this review we present an overview of specific whole-cell bioreporters for heavy metals. A discussion of heavy metal resistance/homeostasis mechanisms is also included since these properties usually form the basis of heavy-metal sensing in bioreporter cells. We make no attempt to compile a comprehensive summary of individual metal-specific bioreporters that have been constructed, as this aspect has been extensively covered in several excellent, recent reviews [7–9]. Instead, our aim is to express our views on the theory and application of metal specific bioreporters and identify some directions for future research and development. We also make an effort to emphasize several aspects of whole-cell bioreporters that, although important, have received relatively little attention.

For the sake of clarity we have used the word "metals" whenever writing about any form of metal. However, it should be noted that heavy metal bioreporters detect only ionic forms of metal; therefore, throughout the text the true meaning of the word "metals" is "metal ions."

1.1 *Principle of Whole-Cell Metal Bioreporters*

Typically, specific whole-cell bioreporters combine a sensor element for the substance of interest and a reporter gene that encodes an easily detectable protein. The sensor element is responsible for recognizing the presence of an analyte. In case of metal bioreporters the sensor element consists of transcription factor gene and respective operator area on DNA fused to a promoterless reporter gene that codes for a protein that produces an easily measurable signal (Fig. 1). Binding of a metal ion to the transcription factor activates the transcription factor and stimulates expression of the reporter gene, thus converting metal sensing into a detectable output signal. The sensing ability of the transcription factor determines bioreporter specificity (which metals are detected) and sensitivity (what concentrations are detectable).

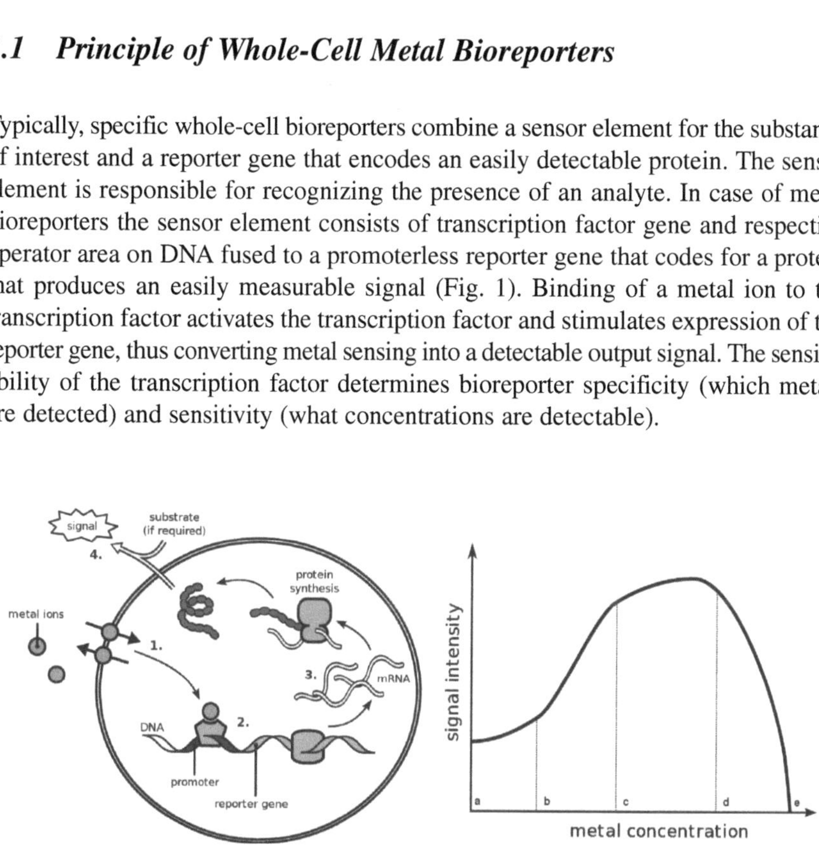

Fig. 1 **a** Principle of whole-cell metal bioreporters. 1. Uptake and efflux mechanisms regulate intracellular metal level; 2. upon binding of the metal ion to the transcription factor, transcription is initiated from the promoter, which is fused to the reporter gene; 3. mRNA for the reporter protein is produced and subsequently translated into protein; 4. depending on the reporter protein, a signal can be measured directly or after the addition of a substrate. **b** Typical output of a whole-cell bioreporter. When the metal concentration is too low to activate the sensor element, only a background signal is produced (a–b). Starting from a threshold concentration (b), the signal increases linearly with increasing concentrations of metal (b–c). A linear regression in this range is used in calculations. At a certain concentration (c–d), the reporter is maximally induced and no further increase in signal can be observed. At a metal concentration high enough to exert a cytotoxic effect (d), the signal rapidly drops to zero (d–e)

The transcription factor is usually derived from the metal resistance system, ensuring that the synthesis of the reporter protein is only directed by the presence of a metal ion capable of binding to the transcription factor. Metal resistance mechanisms are tightly regulated, and are invoked only in the presence of a specific metal in the growth environment. Under natural conditions, exposure to the metal promotes the expression of resistance-mechanism components that enable the cell to defend itself. Bioreporter constructs exploit the cell's natural defense mechanism, redirecting the outcome of transcription factor-metal ion binding into expression of the reporter gene. Because heavy metals are often toxic to cells at subnanomolar concentrations, cells are typically able to detect very low levels of metals. Thus, whole-cell bioreporters with detection limits in the nanomolar range can be constructed for some metals. The most interesting promoters for environmental analysis are found in bacteria that survive in extreme environments contaminated by heavy metals. However, no specific transcriptional regulator for every compound of environmental interest exists. Thus, the range of truly specific bioreporters is limited.

1.2 Reporter Genes in Bioreporters

Reporter genes used in bioreporters should encode proteins that are easy to detect in small quantities, preferably without requiring addition of any reagents. Moreover, it should be possible to detect reporters using diverse instrumentation so that different measurement systems can be employed without having to modify the bioreporters. Reporter genes that best fulfil these requirements are the luciferases and fluorescent proteins. Both bacterial [10] and insect [11] luciferases are detected by measuring chemiluminescence produced in a reaction catalyzed by the enzymes. Bacterial luciferase is a heterodimer comprised of LuxA and LuxB encoded by *luxA* and *luxB*, respectively. These genes are part of the *lux* operon, which consists of five genes. The other three genes, *luxE, luxC*, and *luxD*, encode proteins that form a fatty acid reductase complex that catalyzes the synthesis of aldehydes from fatty acids. These aldehydes are substrates for the luciferase enzyme; therefore, expression of the entire *lux* operon (*luxCDABE*) confers a luminescent phenotype on living bacterial cells without the addition of an external substrate. Limiting expression from the *lux* operon to only *luxAB* requires the addition of an aldehyde substrate, a maneuver that can be technically difficult due to the high vapor pressure of the aldehyde [12]. Insect luciferase is a single polypeptide encoded by the *luc* gene. Luciferase uses ATP as one of its substrates; thus, luminescence is emitted only by living cells. The application of insect luciferase as a reporter protein requires addition of the substrate, d-luciferin, to generate a signal. In contrast, fluorescent proteins, such as green fluorescent protein (GFP) from *Aequorea victoria*, are intrinsically fluorescent once produced and correctly folded [13], and do not require cellular metabolic activity. Hence, even dead cells could be fluorescent, thereby potentially compromising test results. In practice, bacterial luciferase (*lux* operon) is the most versatile choice for bacteria-based bioreporters, whereas bioreporters constructed using insect luciferase (*luc* gene) are best for metal-detecting systems based on eukaryotic cells. In each case, these enzymes exhibit negligible cellular background and thus allow for sensitive detection, and provide for the possibility of using a large variety of detection equipment. Available instruments include multiwell plate readers capable of measuring thousands of samples per day, hand-held single tube luminometers, simple devices using Polaroid film [14] and various fiber optic systems [15]. The properties and applications of these reporter genes in bioreporters have been reviewed previously [16] and are thus not covered in detail here.

1.3 Applications of Bioreporters

Bacterial bioreporters can be used as an analytical means to measure quantitatively heavy metals in samples of interest. In contrast to chemical methods, whole-cell bioreporters do not measure the total amount of metal present in the environment; rather they measure bioavailable metal – the biologically relevant fraction that influences the cell and is capable of passing through cellular membranes.

The most obvious use of metal bioreporters is to measure the bioavailability of metals in different environmental samples. This relatively simple protocol, which requires only mixing of the sample with bioreporter cells, is further simplified by the ability to store bioreporter cells through lyophilization or drying, thus eliminating the need for routine culture and minimizing variation between different batches of cells. Lyophilization or drying of the cells is an important phase in the development of a bioreporter, but is quite often overlooked in reports describing new bioreporters. In addition to direct mixing, bioreporters can be used as part of biosensors in fiber-optic devices [15].

2 Genetics of Metal Homeostasis Regulation in Microbes: Source of Specificity and Sensitivity, and Application in Bioreporters

It is the sensing element that determines the sensitivity (detectable concentrations) and specificity (detected metals) of the bioreporter. Although the binding capacity (metal species and sensed concentration) of the transcription factor used in the sensing element influences sensitivity and specificity, these factors also depend greatly on microbial metal homeostasis/resistance systems. Metal transporters are essential for maintaining homeostasis of essential metals and conferring resistance to toxic heavy metals. Importantly, it is the activity of these transporters that determines the intracellular concentration of metals that is ultimately available for detection by the bioreporters.

2.1 Transport of Metal Ions

Intracellular concentrations of essential trace elements, including cobalt (Co), copper (Cu), nickel (Ni), zinc (Zn), and iron (Fe), must be maintained to avoid metal deprivation. However, in excess, these same metals can have significant toxicity and act as potent disruptors of biological systems. Consequently, the existence of homeostatic mechanisms that finely adjust intracellular concentrations of micronutrient metals within a very narrow range is a fundamental property of all living cells. In prokaryotes, the major mechanisms that maintain cellular concentrations are limited to the highly regulated processes of metal import and export across the cytoplasmic membrane, and metal ion sequestration by metallo-chaperones [17]. This regulation is achieved through specific transcription factors, which sense excess concentrations of metal ions in the cell. The first step in this regulatory process is suppression of the expression of genes involved in uptake mechanisms. However, because this aspect of the process leads to a decrease in gene expression, these transcription factors are not suitable for bioreporter construction. Further increases in intracellular metal concentration activate mechanisms to export the excess metal from the cell. The expression of efflux transporters is induced only if a certain intracellular concentration of the metal is reached and sensed by specific transcription factors. These metal-binding transcription factors form the basis of most sensor elements in bioreporters. The activating concentrations of essential metals are not extremely low, a feature of the bacterial response mechanism that has evolved to avoid metal deprivation. While appropriate from a metal–micronutrient–homeostasis standpoint, this feature sets practical limits on the use of natural transcription factors in bioreporters to detect essential metals.

In contrast, the homeostatic response of prokaryotes to purely toxic metals, such as lead (Pb), cadmium (Cd), mercury (Hg), and arsenic (As), is simple – these must be quickly and efficiently eliminated from the cell to avoid toxicity. Rapid elimination of toxic ions from the cell presumes the ability to detect very low intracellular concentrations; thus, bioreporters with nanomolar detection ranges can be constructed using transcription factors that regulate toxic metal resistance mechanisms.

Different bacterial species have very different transport/resistance systems. Some bacteria may contain several resistance mechanisms for one metal, whereas others may completely lack resistance to the same metal [18]. The mechanisms by which metal ions enter the cell may also differ between bacterial species. All these factors may influence the sensing ability of the bioreporter. Mechanisms that extrude, degrade, or modify the analyte may also compete with its detection, and thus lead to a less sensitive response [19].

2.2 Bacterial Receptors for Metals

Heavy metal resistance systems are not essential for survival; however, in a metal-polluted environment they provide a great advantage for microbial growth. Because the synthesis of proteins is energetically costly, the expression of resistance systems is tightly regulated to prevent inappropriate activation in a clean environment so as to minimize the associated metabolic burden to the host. The expression of most heavy metal transport systems is controlled at the level of transcription. Thus, the transcription of transporters is induced only in the presence of substrate metals in the growth environment. Generally speaking, bacterial toxic-ion resistance systems are determined by clusters of contiguous genes that function as operons, which are regulated by protein binding to upstream operator DNA regions [20]. Three protein classes of metal-responding transcriptional regulators have been described in detail:

1. Members of the MerR family of homodimeric activator proteins usually bind to RNA polymerase-binding operator regions regardless of the presence of a bound metal ion. Upon activation by metal binding, the DNA bends and twists into a position that opens the DNA structure and allows mRNA synthesis to initiate [21, 22].
2. The ArsR/SmtB family of metal-responding regulators consists of repressors that act negatively by binding to the operator region to prevent RNA polymerase binding. When the metal ion binds to the repressor, it is released from the DNA, allowing transcription to occur [20].
3. The third class of transcriptional regulatory proteins form a two-component system comprised of a membrane sensor and an intracellular regulatory protein. In this so-called RS regulatory mechanism, transmembrane histidine kinase membrane sensor (S) autophosphorylates in response to the extracellular stimulus (i.e., upon binding of metal ion). The phosphate is then transferred to a second regulatory protein (R-responder), which is then either activated or inactivated, thereby achieving transcriptional control through positive or negative regulation of DNA-binding [20]. Unlike the first two transcription–regulation mechanisms, which detect cytoplasmic metal, RS regulatory mechanisms detect extracellular or periplasmic metal. RS transcription systems are still incompletely understood from a functional, mechanistic standpoint and are very seldom used in bioreporters.

The regulatory protein used in a bioreporter construct determines the effective detection range for a given metal. Due to the limited possibilities for coordinate binding between proteins and metal ions, metal-activated transcription factors usually sense several different heavy metal ions with similar properties (see below). Thus, it is generally not possible in practice to construct a bioreporter that senses only one metal. However, because binding affinities vary between metals, different metals are sensed at different concentrations.

2.3 *Resistance/Homeostasis Mechanisms for Specific Metals and Their Use in Bacterial Bioreporters*

All bacteria have defense mechanisms against metals. These mechanisms exist for almost all metals that are present in the environment in concentrations sufficient to become toxic to bacterial cells.

Although, in theory, metals can be divided into essential metals, which have specific uptake and efflux mechanisms, and nonessential metals, which have only specific efflux mechanisms, in practice the difference between metals is not that clear cut. For example, Zn uptake mechanisms incidentally also import Cd and Pb. The reverse is also true; in fact, Zn efflux mechanisms export Cd and Pb even more efficiently than Zn itself. This broad specificity of resistance/homeostasis mechanisms has implications for bioreporters, as discussed in detail for each metal in the following sections.

2.3.1 Zn, Cd, Pb

Zn is an essential metal that is actively imported to satisfy cellular Zn requirements; in *Escherichia coli* alone there are four different mechanisms for Zn uptake [20]. Zn serves as a cofactor for more than 300 enzymes and is required to maintain the structural stability of macromolecules. However, in excess, it inhibits the aerobic respiratory chain, and can have significant toxicity. Cd and Pb, the chemical "look alikes" for Zn, have no known beneficial role in cells. Cd and Pb enter bacterial cells via transport systems for essential divalent cations, such as Mn^{2+} [23, 24] and Zn^{2+} [25–27].

Cellular extrusion mechanisms do not substantially differentiate between Zn, Cd, and Pb. In many cases, Zn-resistance determinants are genetically indistinguishable from those conferring resistance to Cd and Pb. For example, the divalent metal efflux P-type ATPases, ZntA (*E.coli*), CadA (*Staphylococcus aureus*) and CadA (*Pseudomonas putida*), all efficiently export Pb, Cd, and Zn [28–32]. Perhaps not surprisingly, expression of these metal transporters is similarly regulated by several metals. The transcription factor, ZntR, which acts on the *zntA* promoter, is activated by Hg, Cd, Pb, and Zn [33, 34]; expression from the $cadA_{\mathrm{P.putida}}$ promoter is induced by Zn, Cd, Pb, Co, and Hg [31, 32], and the $cadA_{\mathrm{P.aeruginosa}}$ promoter is activated by Zn, Cd, Hg, Ni, Co, and Cu [35]. All the aforementioned transcription factors belong to the MerR family. The situation is similar for other types of transcription factors. CadC, a member of the ArsR family that regulates $\mathrm{CadA}_{S.aureus}$ synthesis, is derepressed by Cd, bismuth (Bi), Pb, Co, and Zn [36]. Thus, the range of inducers is actually much broader than the range of substrates recognized by the transporter. This property of coinducibility creates constraints on bioreporter design; accordingly, no bioreporters that are individually specific for Zn, Cd, or Pb have been developed.

Bioreporters constructed using ZntR and the promoter of the *zntA* gene are most sensitive to Cd, and Pb, but they are also induced by higher concentrations of Zn and Hg [37–39]. Similarly, bioreporters that use CadC and the promoter for *cadA* from *S. aureus* are not specific for Cd, but are coinduced by Cd, Pb and even antimony (Sb) [40–42]. Some differences in metal specificities have been noted between different strains [40], but a specific sensor for one metal does not exist.

2.3.2 Cu, Ag

Cu is an essential trace element, acting as a cofactor for several enzymes. However, it can also cause serious cell damage through formation of free radicals. Ag(I) is chemically very similar to Cu; however, it cannot replace Cu in enzymes and exerts only toxic effects.

Very little is known about how Cu enters the cell, although it appears to cross the outer membrane through porins. How it crosses the cytoplasmic membrane has not been conclusively demonstrated in bacteria beyond the possibility that Cu(I) may equilibrate across this barrier [43]. The best characterized protein with a suspected Cu- and Ag-uptake function is CopA from *Enterococcus hirae*; however, its purported function has been modeled based only on indirect evidence [44].

Similar to divalent metal transporters, the CBA transporters and P-type ATPases for monovalent metals are not specific to one ion. They fail to differentiate between Cu(I) and Ag(I) and thus transport both [45, 46]. The transcription factor, CueR, is a member of the ArsR family that regulates $copA_{\mathrm{E.coli}}$ expression; it derepresses the *copA* promoter in the presence of either Cu or Ag [47], but gold (Au) is also effective [48]. Ag ions are detected at much lower concentrations than Cu [46]. Such a regulatory mechanism makes biological sense considering that Ag is purely toxic, but Cu is needed for cellular processes. Eukaryotic systems are similar in this regard. Ace1, which induces the synthesis of the Cu detoxification metallothionein, Cup1, in *Saccharomyces cerevisiae* is activated by both Cu and Ag [49]. Interestingly, Cd, Co, and Zn also bind to Ace1, but do not induce synthesis from the respective promoter [50]; they thus act as competitive inhibitors for Cu and Ag binding [49].

Different Cu-responsive elements have been applied in the development of Cu bioreporters. The systems that have been constructed and calibrated include those from *Cupriavidus metallidurans* [51] and *Pseudomonas fluorescens* [52], both combined with a *lux* reporter, and *E. coli* [53] and *S. cerevisiae* [49, 54], constructed using *luc*. The Cu-responsive bioreporters usually also respond to Ag, as predicted from the induction profiles of Cu-related genes.

2.3.3 As

Certain prokaryotes use As oxyanions for energy generation, either oxidizing arsenite or respiring arsenate [55]. Microbes that do not gain energy from arsenate reduction use reduction instead as a means of coping with high arsenic concentration in their environment. Organisms take up As(V) via phosphate transporters, and use aquaglyceroporins, which normally transport organic polyols, to import As(III) and Bi(III) [56].

Resistance to arsenicals (As[III] and As[V]) and antimonials (Sb[III]) is achieved through the action of *ars* operon-encoded proteins [57, 58]. Arsenate that has entered the cytoplasm is converted to As(III) and expelled from the cell by a transporter. Expression of the *ars* operon is regulated by the As(III)/Sb(III)-responsive transcriptional repressor, ArsR [59], which is also induced by Bi(III) in *S. aureus* [57, 58].

Bioreporters for As have been constructed based on *lux* [60], *luc* [61, 62], and fluorescent proteins [63] using ArsR from either *E. coli* [59] or *S. aureus* [57]. The rank order of metal specificity of *E. coli* As bioreporters is As(III) > As(V) >Sb(III) >>Cd(II) [62], whereas that for *S. aureus* As bioreporters is Sb(III) ≈As(III)

>Cd(II) >>As(V) [61]. In both cases, the detectable concentration of As(III) is in the high nanomolar range.

2.3.4 Cr

In the biological environment, chromium (Cr) exists mainly in the form of an oxyanion (CrO_4^{2-}). Chromate enters the cell through the sulfate uptake pathway, reflecting the analogous chemical structures of these two oxyanions [64]. Chromate resistance is based on the active extrusion of chromate anion or reduction of Cr(VI) to the less toxic Cr(III) by the action of various enzymatic or nonenzymatic processes. The reductases that act on Cr(VI) are rather nonspecific, having chromate reduction ability as a secondary function [64]. Export of chromate from cells is conferred by the action of a ChrA transporter, which has been identified in *Ochrobactrum tritici* [65], *Pseudomonas aeruginosa* [66, 67], and *C. metallidurans* [68]. The expression of *chrA* has been found to be inducible by chromates, dichromates and chromium, but not by other oxyanions or metals [65, 69, 70] making Cr sensing relatively specific. This property has enabled the development of bioreporters for Cr that selectively detect chromate oxyanions over an environmentally relevant concentration range [39, 69].

2.3.5 Ni and Co

Ni and Co are important cofactors for several enzymes. Entry of these ions into the cell is facilitated by specific uptake mechanisms [71] or by nonspecific permeases [72].

Resistance to Ni and Co can be encoded by the same determinant, as exemplified by *cnrCBA* of *C. metallidurans* [73] and *rcnA* of *E. coli* [74]. A similar operon (*ncc*) is found in *Alcaligenes xylosoxidans* 31A, where it additionally codes for Cd resistance [75]. In each case, expression of these systems is regulated by both Ni and Co [74, 76]. However, these metals can be detoxified separately. Co resistance is achieved through a *czcCBA* determinant that is responsible for Zn, Cd, and Co efflux in *C. metallidurans* CH34 [77]. *Synechococcus* also encodes a specific Co transporter whose expression is induced by Co and Zn [78]. Expression of the *nrs* operon of *Synechococcus*, on the other hand, is induced by both Ni and Co, but it only confers resistance to Ni [78]; NreB from *Achromobacter xylosoxidans* 31A is a specific Ni exporter that is also specifically induced by Ni [79].

The only bioreporter that has been constructed for Ni detection is based on the *cnr* operon from *C. metallidurans* [80]. It is inducible by both Ni and Co, with Ni being detected at 100-fold lower concentrations.

2.3.6 Hg

Among toxic heavy metals, the mercuric ion stands alone in that it has its own uptake system as part of a resistance mechanism that allows efficient detoxification of Hg ions. Unlike other metal resistance systems, the Hg resistance system encoded by the *mer* operon is truly specific. The resistance mechanism is based on the transformation of ionic Hg to metallic Hg, which evaporates from the cell [81]. Because this process is impossible for other metals, its only substrate is Hg. Resistance to the Hg^{2+} ion alone or to Hg^{2+} ions together with organic Hg compounds is encoded by different *mer* operons, defined as narrow-spectrum or broad-spectrum operons, respectively [81]. The resistance to organic compounds is mediated by organomercurial lyase (MerB), which catalyses the breakdown of the carbon–Hg bond in organic compounds to generate an Hg^{2+} ion and the organic residue. MerR, the transcription factor that regulates the expression of the *mer* operon, is relatively specific for Hg, but is also activated by Zn and Cd at much higher concentrations [82].

A vast number of bioreporters for Hg has been developed, arguably because the structure and function of the *mer* operon have been characterized in such detail. Most of the bioreporters have been based on the narrow spectrum MerR and are therefore restricted to the detection of Hg^{2+} ions [83, 84, 85]. The detection of organomercurial compounds is also of great interest in most applications. This can be accomplished using components of the broad-spectrum *mer* operon since MerR from these operons recognizes some organomercurials directly [86]; expressing the *merB* gene in a bioreporter further enhances the breadth of the organomercurial detection spectrum to include monomethylmercury and dimethylmercury [87].

2.3.7 Fe

Fe bioreporters have been constructed primarily for testing the availability of dissolved Fe to phytoplankton rather than for detecting Fe contamination. Therefore, the sensor systems used for Fe bioreporters differ from those used to detect other metals. Fe bioreporters have typically used promoters that are activated by Fe deficiency – or more precisely, turned off when a sufficient amount of Fe is present. Such systems include Fe-uptake mechanisms that are induced in a low-concentration environment to maximize Fe retrieval. The best known cellular factor associated with modulating intracellular Fe levels is the ferric uptake regulator (Fur), which regulates expression of several genes involved mainly, but not exclusively, in Fe homeostasis [88]. Fur acts as a transcriptional repressor of Fe-regulated promoters by virtue of its Fe-dependent DNA binding activity. For example, Fur regulates the expression of genes for the synthesis of siderophores, which are essential for the uptake of Fe(III) by bacteria [89]. At high concentrations, Co^{2+} and Mn^{2+} ions presumably mimic Fe^{2+} [90, 91], and Zn and Cd can also activate Fur in vitro [90]. Therefore, these metals could possibly interfere with Fe detection.

Fe bioreporters have mainly been constructed in cyanobacteria [92–94] although *Pseudomonas* sp have also been used [95, 96], mainly for the detection of available Fe as a nutrient.

2.4 Improvement of Sensitivity and Specificity

Dozens of bioreporters have been constructed and their suitability for testing environmental samples has been demonstrated. Despite this, bioreporters are not yet widely used, partly due to low sensitivity and lack of specificity. The amount of bioavailable metal in complex matrices can be <1%, even in highly contaminated samples. Reliable detection of such low concentrations requires improved detection limits. Also, the ability to characterize specifically the contaminants in a sample requires bioreporters that respond to only one metal.

The main components of a bioreporter that influence sensitivity and specificity are the sensor element and the bacterial metal homeostasis system. Some strategies for improving the sensitivity and/or specificity of bacterial bioreporters include (1) enhancing the import of contaminant into the cell, (2) hindering the export of detectable ions from the cell, and (3) modifying the sensing element.

Almost all bioreporters constructed to date detect only intracellular metal; thus, metal ions must be transported into the cell to be detected. Different bacterial species possess different mechanisms for metal uptake, resulting in organism-specific differences in the amounts of bioavailable contaminants that can be measured. To determine all potentially bioavailable contaminants, bioreporter organisms could be genetically engineered to make their membrane more permeable for the contaminant or to contain additional uptake mechanisms. Such a strategy has been tested on the Hg bioreporter [84]. In addition to detoxification proteins, the Hg-resistance operon also encodes proteins that mediate the uptake of Hg^{2+} ions into the cell. Increased uptake allows metal ions to accumulate in the cell and reach intracellular metal concentrations suitable for induction under lower extracellular Hg concentrations than would be possible if only passive Hg uptake mechanisms were involved. When uptake proteins are present in the cell, lower extracellular Hg concentrations are needed to induce expression from the Hg-inducible *mer* promoter [84, 97], thus improving Hg sensitivity in bioreporter applications. Although specific uptake systems only exist for a limited number of metals, the proteins responsible for uptake of most metals are known (see above). Overexpression of these systems in bioreporter strains would thus be one approach to improving sensitivity.

Another way to achieve more rapid accumulation of metal ions in the cell is to eliminate efflux transporters. Most environmental bacteria contain some type of efflux transporters for heavy metals [18]. Efflux mechanisms compete with the detection of metal ions and thus lead to a less sensitive response. If all the metal ions entering the cell were immediately removed, no binding to transcription factors would occur and hence no signal would be generated. However, in the absence of efflux transporters, metal ions start accumulating immediately in the cell; when the threshold concentration is reached, expression of the reporter gene is induced. Indeed, mutations of heavy metal transporters have, in some cases, proved useful in improving the sensitivity of bioreporters. Removal of CopA, a Cu/Ag-transporting ATPase, from *E. coli* rendered a bioreporter employing the CueR–PcopA sensor system 15-fold more sensitive to Cu and 8-fold more sensitive to Ag [46]. Similarly, deletion of *zntA*, the gene for a Zn/Cd/Pb-translocating ATPase, improved the detection sensitivity for Zn by about three orders of magnitude, while increasing Pb and Cd sensitivity by 15-fold and 2-fold, respectively [30]. Removal of an efflux system, however, does not always improve bioreporter detection limits. Tauriainen et al. [62] compared the performance of two *E. coli* As bioreporters – one with an intact As detoxification system and another containing a deletion in the chromosomal arsenite-resistance operon. They found no differences in performance of these bioreporters, showing that the lack of the *ars* operon did not cause a remarkable accumulation of As in the cell.

Although increasing sensitivity is desirable, if pushed to an extreme it would be possible to make cells too sensitive, resulting in near-continuous induction of bioreporter by (for example) trace amounts of metal in the growth medium. In fact, trace amounts of heavy metals in microbiological/analytical grade chemicals can be enough to induce maximally a highly sensitive bioreporter, eliminating the possibility of detecting additional metal in environmental samples (Hynninen et al., unpublished results). Because bioreporters are meant for rapid and inexpensive analysis of environmental samples, replacing microbiological/analytical grade with highly purified chemicals would be impractical.

Although the addition of influx transporters or deletion of efflux transporters might also affect the specificity of the bioreporter (depending on the substrate range of the transporter and sensing system), the most direct way to influence bioreporter specificity is through the sensor element. Several successful mutations have been introduced in transcription factors for organic compounds that have changed the substrate range [98–100]. However, changing the effector specificity of metal-binding transcription factors has not been a trivial task. Desilva et al. [101] have shown that metal specificity can be changed by altering the relative location of cysteine residues in metal-binding proteins. Single amino acid mutations in CueR, the Cu/Ag/Au-responsive transcription factor, abolished its response to Ag and greatly increased its response to Au, without altering Cu responsiveness [48]. Similarly, mutations in ZntR, the Zn/Cd/Pb-responsive transcription factor, yielded proteins with different response profiles than wild-type ZntR [102]. However, this mutagenesis strategy yielded no mutants that were specific for only one metal. Mutations in the Hg/Cd/Zn-responsive transcriptional regulator, MerR, rendered it more sensitive to Cd, without abolishing responsiveness to Hg [103]. Such mutants with changed metal specificity have not yet been used in the construction of bioreporters, but they could represent an important component of bioreporters in the near future.

3 Practical Testing Issues

The motivation for the development of whole-cell metal bioreporters is usually the detection of metals in different matrices. To reach that in relatively difficult matrices, such as soils, sediments, biological tissues, and food, the assay protocol plays a major role. In this section, issues influencing the performance of metal bioreporters will be discussed.

3.1 The Effect of Test Conditions on the Performance of Whole-Cell Metal Bioreporters

Test conditions play a pivotal role in the analysis of heavy metal by bacterial bioreporters. For example, the test environment and duration must be such that the reporter protein can be synthesized, but it should not cause complexation, changes in the association–dissociation kinetics between the matrix and the metal, precipitation of metals, or other effects that would interfere with heavy metal speciation. This is not an insignificant consideration since most heavy metals that are the target of constructed bioreporters belong to the soft metals, which form relatively strong complexes with different ligands [104].

Physiological status of the cells considerably affects the performance of the whole-cell metal bioreporters. Cells that are in the exponential growth phase have often been used in bioreporter assays, as this has proved to be optimal for the performance of bacterial cells. For example, the best performance of the *luxCD-ABE*- expressing Cu and Zn reporters (*E. coli*) in cells grown in rich LB media was obtained at early exponential growth [38]. That study also showed that the response of the bioreporters decreased at higher cell densities, most likely due to the absorption of metals by cell wall components or to light absorption due to the turbidity of the bacterial suspension. A similar effect has been shown with bioluminescent Hg-sensing *E. coli, P. putida*, and *Enterobacter aerogenes*, where reducing cell density by two orders of magnitude increased the sensitivity of the assay by two orders of magnitude [105, 106]. In the case of *Bacillus subtilis*, the lowest concentrations of the target metals were sensed in the early stages of growth when cell density was low [61]. The strong dependence of the expression from heavy metal-inducible promoters on the growth phase has been shown in *P. putida* where three metal-responsive promoters were highly induced in the exponential phase, but not in the lag or stationary phases [32].

The chemical nature of the test medium is a major contributor to the bioavailability. Depending on specific cellular requirements, microorganisms are usually cultured in either complex media containing (for example) yeast extract and amino acids, or in defined media. Unfortunately it is not possible to use one defined medium for all bioavailability tests since different organisms have different requirements. Both rich, complex media and mineral-defined media have their own benefits and drawbacks. Since the presence of nutrients is essential for the function of living cells, especially for cells deficient in certain biosynthetic pathways, the use of amino acid-supplemented rich media could be considered most generally suitable. Since bioreporters are based on the biosynthesis of reporter protein(s), the addition of amino acids may enhance the response of the whole-cell metal bioreporters. If the reporter protein is an enzyme, the metabolic burden for the cells can be considerable. For example, the expression of bacterial luciferase operon may use up to 10% of the cell's total energy output since the luminescence is produced all the time [107]. However, it is also important to take into account the fact that components of the test medium may affect metal complexation (reviewed in [104]). The concentrations of organic ligands (and some inorganic ligands) are considered to be the main factors that influence metal speciation in test media and it has been demonstrated that the components of complex media may bind metal ions, and thereby reduce the toxicity of these metals [108, 109]. There is also a clear correlation between the amount of dissolved organic carbon (DOC) and the bioavailability of Hg as detected with a luminescent Hg-bioreporter: in the presence of higher concentrations of DOC, the response of the bioreporter decreases significantly [110]. Therefore, when possible, rich media should be avoided, and bioavailability tests should be carried out under conditions in which the levels of possible metal-binding ligands have been minimized. In some cases, however, the inducibility of metal-regulated promoters by certain metals could be better in a rich medium [32]; thus, it is necessary to test different media during assay optimization.

Among the most important inorganic ligands with respect to impact on metal speciation are phosphates and high levels of chloride [104]. Phosphates have a strong complexing effect on heavy metals. Phosphates are usually required both for their buffering capacity and their value as a nutrient for bacterial growth. Therefore, media used in heavy metal bioreporter assays usually contain a high level of phosphate or phosphate-containing buffers. For buffering purposes, phosphates could be substituted by morpholinepropanesulfonic acid (MOPS), or other sulfonic acid-based buffers, such as MES, PIPES, TES, or HEPES [104]. β-Glycerophosphate has been suggested as a nutrient substitute for inorganic phosphate [111]. Indeed, detection of Cu and Zn by luminescent reporter bacteria proved to be more sensitive in a glycerophosphate-containing medium than in a rich LB medium [38].

In addition to phosphates, the content of NaCl in the test medium also influences the speciation and bioavailability of heavy metals. High levels of NaCl (2–4 M) have been shown to reduce the toxicity of $CdCl_2$ by about 10%, most likely due to the formation of anionic chloro-complexes [112]. Similarly, a nearly linear decrease in Hg bioavailability has been shown to occur as NaCl concentrations are increased up to 0.1 M [110].

It should be noted that, although nearly all studies report the calibration of whole-cell bioreporters, it is difficult to evaluate the effects of test media because additional parameters, including the reporter cell, temperature and incubation time, also typically vary. Methods for modeling the speciation of metals in a given test medium are available if the appropriate dissociation–association constants are known. Computer modeling of metal speciation could be performed in parallel with toxicity or bioavailability tests [113]. The computer programs used most often to calculate metal speciation are MINEQL or its derivatives (e.g., MINTEQA). While studies have shown that it is possible to correlate the predicted free ion concentration with biological effects, these findings are not without controversy. For example, changes in Cu bioavailability in the presence of EDTA coincided with modeled changes in Cu activity, indicating that Cu–EDTA complexes were not bioavailable to the Cu specific bioreporter [114]. In contrast, changes in Cu bioavailability in the presence of citrate did not correspond to changes in the concentration of Cu^{2+} ions, indicating that Cu–citrate complexes were fully bioavailable to the reporter strain. Finally, the response of the Cu-reporter strain to Cu in the presence of dissolved organic matter (DOM) indicated that Cu formed both bioavailable and biologically unavailable complexes with DOM [114]. For Hg, a negative correlation between bioavailability and modeled concentration of negatively charged forms of mercuric chloride has been reported, showing that negative mercuric chloride ions are not available to bacteria [110].

Test media optimized for whole-cell metal analyses also determines the pH of the test environment. Because a pH in the neutral range is optimal for living cells, most microbial growth media are well buffered to approximately pH 7. However, dilution of growth-media-suspended bacterial cells with the environmental sample (e.g., water, aqueous soil extract or soil suspension) may change the pH of the final test sample. Because metal speciation is greatly affected by pH, this change in pH could alter the apparent bioavailability of metals. For example, under anaerobic conditions, 1 ng L^{-1} Hg did not evoke a response from *luxCDABE*-based Hg reporter bacteria at or above pH 6, but induced a remarkable (up to 100-fold) induction of bioluminescence in bioreporter cells at approximately pH 5 [115]. Under the anaerobic conditions used in this experiment, the speciation of Hg at pH 5 apparently favored the formation of Hg species that were more bioavailable to bacterial cells. Similar results have been obtained in the presence of dissolved organic carbon in the test environment in which remarkably more Hg was bioavailable to luminescent sensor bacteria at pH 5 than at pH 7 [110]. This change in bioavailability may be due to an apparent effect of pH on a facilitated mechanism that mediates Hg(II)-uptake by the cells [116]. Similarly, Cd speciation has been found to greatly depend on pH. However, this metal tended to form the most bioavailable species at pH 6 [115].

In applications involving eukaryotic luciferase, which requires a luciferin substrate, the pH must be lowered to the slightly acidic conditions that favor the transfer of luciferin across cell membranes [85, 117]. After the addition of d-luciferin, microbial cells are usually allowed to incubate for about 20 min [39, 87, 118–120]. Given the well-known pH-dependence of heavy metal speciation, even this length of time may allow the pH to drop enough to influence the outcome. Thus, as suggested by [121], the composition and pH of the growth/maintenance medium should be critically considered and, when presenting and analyzing the data, the results obtained need to be checked carefully for unintended pH effects.

Incubation time of bioreporter with the sample depends greatly on the reporter element used. The time required for the induction reflects the time required for expression of reporter gene, and synthesis of sufficient reporter protein to allow emission of a measurable signal. A comparison of four of the most commonly used reporters – eukaryotic luciferase *luc*, the *luxCDABE* bacterial bioluminescence cassette, and fluorescent protein-encoding genes, *gfp* and *dsRed* – showed that a 30-min incubation was enough to induce *luc* and *luxCDABE* in Hg and As bioreporters. In contrast, at least 2 h was required for the induction of *gfp* and 8 h was needed to induce *dsRed* [122]. The long incubation times required for fluorescent proteins are a consequence of their slow folding; for example, DsRed requires days to yield a maximal fluorescence signal at room temperature [123]. However, these limitations may no longer apply as these early fluorescent protein variants have been largely supplanted by a new generation of fluorescent proteins with faster maturation kinetics [124, 125].

Experiments with bioluminescent whole-cell bioreporters have employed incubation times ranging from 30 min to 6 h. Time course experiments showed that maximal induction for *E. coli luxCDABE*-based Zn and Cu reporters occurred between 60 and 160 min of incubation [38], whereas incubation times of 4 and 5 h were optimal for *P. putida* and *E. aerogenes*, respectively [106].

The test temperature may also influence the performance of whole-cell bioreporters. Usually, the temperature that is optimal for growth of the host cell is used for testing, as it is normally also optimal for inducibility. The inducibility of *luc*-expressing As and Hg bioreporters in *E. coli* has been shown to be optimal at 37 °C, whereas room temperature is optimal for the induction of these bioreporter systems in *P. fluorescens* [118]. In the case of Hg-inducible *P. putida* and *E. aerogenes*, which both grow optimally at 28 °C, the highest luminescence was measured at 35 °C and 37 °C, respectively [106]. However, this study tested only a single Hg concentration, so the effect of higher temperature on the overall performance of these bioreporters remains unclear.

The reporter system may also set some restrictions on the test temperature. Widely used *Vibrio fisheri luxCDABE* bacterial bioluminescence system becomes labile at temperatures above 30 °C [16]; thus, a lower temperature should be used in bioreporters applying this operon. As an alternative, the *lux*-operon from terrestrial *Photorhabdus luminescence* could be used, as the enzymes encoded by this operon are functional at temperatures as high as 45 °C [10].

3.2 *The Effect of Host Cells on the Performance of Whole-Cell Metal Sensors*

Different host bacteria have been used to construct whole-cell heavy metal bioreporters. Different responses can be expected from different hosts, in keeping with the observation that bacterial species exhibit differences in their metal homeostasis/resistance mechanisms. Host cells may also set some restrictions on the test conditions. In addition to host effects on induction temperature optima, noted above, the microbial host may dictate the test media used. Whereas some whole-cell bioreporters are based on wild-type cells able to survive in any mineral media, other bacterial hosts require certain amino acids for physiological activity.

Because its metal resistance systems are so well studied and it is easy to genetically manipulate, *E. coli* has been the most favored host for whole-cell heavy metal bioreporter applications. Other Gram-negative bacteria that have been used as hosts for metal-bioreporters include *C. metallidurans* (formerly *Alcaligenes eutrophus*, *Ralstonia eutropha*, or *Ralstonia metallidurans*) [51, 69, 70, 80, 126], *P. putida* [127] and *P. fluorescens* [52, 106, 118]. Gram-positive bacteria have been used as hosts for metal bioreporters in relatively few studies. Examples include *S. aureus*- and *B. subtilis*-based luminescent bioreporters constructed for the detection of arsenite, antimonite, Cd [40, 61], Zn, and Hg [128].

The bacterial cell that hosts the heavy metal response elements may have a crucial effect on the response of the bioreporter to heavy metals. *S. aureus*- and *B. subtilis*-based bioreporters employing the same receptor have been shown to exhibit more than a 1,000-fold difference in the detection of Pb while exhibiting similar sensitivities for Cd [40]. A comparison of the hosts, *E. coli, P. fluorescens, S. aureus*, and *B. subtilis*, indicated that the response of bacterial bioreporters to Hg, Cd, and Zn was mostly dependent on the sensing (metal-response) elements [128]. This supports the assumption that sensing is tightly connected to the metal-regulated element used in bioreporter construction. In tests performed under identical conditions, a comparison of *E. coli, Vibrio anguillarum*, and *P. fluorescens* Hg bioreporters carrying the same sensing element revealed that the Hg-detection capabilities of *E. coli* and *V. anguillarum* were similar, whereas *P. fluorescens* was less sensitive [129]. Since these tests were performed using identical conditions, the possibility that additional differences might be observed in assays optimized separately for each strain cannot be ruled out.

Although preliminary experiments may show no clear differences between different bacterial species, the host for a bioreporter construct should still be chosen carefully. One factor that should be considered is the expected use of the bioreporter. If soil samples are to be analyzed, a soil bacterium would be a suitable host. Because they are compatible with the use of minimal growth media, environmental strains might be preferable to laboratory strains that lack some biosynthesis pathways.

In addition to bacteria, yeasts have also been used as hosts in some metal bioreporters. *S. cerevisiae* has been used as a host for a Cu bioreporter [49, 54], and *S. cerevisiae* and *Hansenula polymorpha* have been used as hosts for Cd bioreporters [130]. Both yeast Cu and Cd bioreporters are comparable in sensitivity to bacterial Cu and Cd bioreporters. However, in bioavailability assays, somewhat less Cu proved to be available to yeast cells than to bacterial cells [131, 132]. Because yeasts are eukaryotic organisms, their metal homeostasis and detection mechanisms are more complicated and differences in metal bioavailability with bacteria are to be expected.

3.3 *The Effect of Sample Matrix on the Test Results*

The test results may be greatly influenced by color and solid matter of the sample matrix, especially if the tests are based on an optical end-point. Moreover, samples may contain toxic substances that reduce the response of bioreporters. Thus, it is vitally import to include appropriate controls. In general, two main approaches are used to try to overcome interference by the sample matrix: (1) removal of the disturbing matrix from the test environment and (2) use of specially designed bacteria that constitutively produce the measured signal as controls to measure (and ultimately subtract out) the effect on the signal due to the matrix alone.

Unfortunately, samples quite often affect the bioreporters in ways that are not related to the target metal content. Usually these are toxic, inhibitory effects, but other effects, such as nonspecific stimulation, can also take place. Although removing the disturbing matrix from the test may not seem to be a viable solution, this approach has been tried. For example, soil–water extracts have been used in place of the soil itself in the analysis of heavy metals from soils [118, 133–135]. However, the results obtained from an analysis of soil–water extracts do not necessarily reflect the actual bioavailability of the sensed metal; whole-cell bioreporters in direct contact with the soil detect much higher bioavailable concentrations of metal [120]. Another possibility is to remove the sample matrix and its influence on the signal just before the measurement. The method is based on density centrifugation and allows the bacterial response to be measured without any physical interference [9]. Immobilization of the reporter cells in alginate on optical fiber tips has also been used to eliminate the need to take into account the effects of the sample matrix on the resulting signal [15].

The second approach is to use control strains that lack the metal-sensing element, but are otherwise identical to the metal sensor cells. Because such control strains emit the signal (e.g., luminescence, fluorescence) constitutively, any changes in the signal are caused by changes in the cell physiology, and not by changes in the concentration of the target metal, except at concentrations that are overtly toxic. A number of different configurations have been used to construct such control strains. These include the use of (1) a promoterless reporter gene [42], (2) a modified metal-response system (usually the promoter) that is no longer controlled by the metal and instead drives constitutive expression of the reporter system [110], and (3) a constitutive instead of metal-inducible promoter [15, 52, 53, 61, 85, 119, 120, 136].

In a solution that contains only metal, these control strains report on metal toxicity, but in general, the use of constitutive-signal-emitting control strains enables us to take into account the optical effect of the sample matrix on the reporter signal. It also allows us to assess the general effects of the sample on the physiology of the reporter cell. Although it could be assumed that, due to its content of toxicants, the sample usually adversely affects the microbial physiology and decreases the produced signal, the opposite may also occur. An explanatory case is the increase in bioluminescence of a *lux*-based naphthalene bacterial bioreporter due to organic solvents present in the sample [137]. The reason for that erroneous increase was the perturbation of bacterial membranes by the solvents followed by increase in fatty acid synthesis and a subsequent increase in the supply of aldehydes that are substrates of the bioluminescence reaction. This increase in substrate levels accounted for the increased light production. The presence of nutrients in the sample may also increase the viability of cells, especially if the test is carried out in a poor medium, and thus have a false positive effect on the emitted signal.

Despite the relatively frequent use of constitutive-signal-emitting control cells in bioavailability experiments, their response is relatively rarely taken into account in the final calculations. The first report detailing the calculation of correction factors based on the response of control cells was an analysis of the quantitative increase or decrease in the reporter signal as a general response of cells to synthetic water samples [53]. Since then, a similar method has been used in tests to determine the amount of bioavailable metals in environmental samples [15, 120, 131, 132, 136, 138].

3.4 *Interpretation of Results*

As discussed above, whole-cell metal sensors typically register an increase in a reporter signal with increasing metal concentrations. The increase in signal may be presented in arbitrary instrument readings, such as light units in the case of bioluminescence. However, because raw relative values may vary depending on the sensing system and measuring instrument used as well as on test conditions, which may influence microbial physiology, data presented in this way may be difficult to compare between experiments. A more versatile approach that allows accurate comparisons between experiments is to normalize the signal emitted by the tested cells to the background signal (i.e., the signal produced in the absence of inducing analyte). This ratio, which has been called the induction coefficient, or induction-to-noise or signal-to-noise ratio, shows the fold-induction of the signal in the presence of analyte.

Plotting the response of the whole-cell metal bioreporter against known concentrations of metal standards yields a calibration curve (see Fig. 1b). To use whole-cell bioreporters as a bioanalytical device for measuring metal concentrations in unknown samples, the concentration range over which a metal causes a concentration-dependent increase in the reporter signal and the smallest concentration that can be reliably determined (limit of detection) should be defined. In practice, this ideal is often not met; papers presenting real-world applications of whole-cell metal bioreporters do not always show the calculations behind the sensor analysis in detail, and the limits of detection are frequently not reported. These omissions significantly complicate comparisons of bacterial bioreporter results from different studies. In an important paper presenting a thorough description of the calculations that enable the quantification of the bioreporter response, the limit of detection was defined as a value twofold above the background signal plus its standard deviation [53].

4 Applications and Interests

4.1 *Measuring Metals: Chemistry vs Sensor Analysis*

Bacterial bioreporters can be used as an analytical means to measure quantitatively heavy metals in samples of interest. In contrast to chemical methods, whole-cell bioreporters do not measure the total amount of metal present in the environment; rather they measure bioavailable metal – the biologically relevant fraction that influences the cell and is capable of passing through cellular membranes. Bioavailability of chemicals has been defined differently in different disciplines; however, in environmental sciences it is considered to represent the soil-bound chemical accessible for possible toxicity, or the fraction that has entered the cell and become available at a site of biological activity [139].

Accurate quantitative estimation of total metal is only possible in samples where all metal is present in a form that is capable of entering the living cells of the reporter bacteria. If the aqueous medium becomes more complicated – for example, if it contains fine particles or chelators that sorb or bind heavy metals – the amount of metals determined by chemical methods and bioreporters may differ markedly. The decrease in bioavailability of heavy metals due to chelators present in environmental samples has been modeled using different chelating agents. The addition of EDTA, CTDA, or Na_2S to aqueous test samples reduced the bioavailable fraction of Pb and Cd by more than 60% [38, 140]. Cu bioavailability also decreased rapidly upon addition of EDTA [38, 52, 114] or in the presence of DOC [141]. The presence of DOC has also been shown to reduce Hg bioavailability [110]. On the other hand, it has been reported that dissolved organic matter–Cu complexes exhibit partial bioavailability in *P. fluorescens* [114], demonstrating the complexities associated with interpreting results from microbial metal bioreporters. In addition to the above-mentioned chelators, metals in the aqueous phase may precipitate with a number of ligands, including carbonate, hydroxide, silicate and phosphate, resulting in nonbioavailable forms [141].

Interpreting the results from whole-cell metal bioreporters becomes even more difficult if solid samples are analyzed, since solid phase of soils and sediments has a high sorption capacity for heavy metals [142]. For standard chemical analyses, soil samples are typically treated using harsh methods involving nitric acid, aqua regia or a mixture of nitric acid–hydrofluoric acid–perchloric acid to destroy the solid soil matrix [143]. Because whole-cell heavy metal biosensors are able to determine only the cell-entered metal fraction and thus require aqueous-based samples, the analysis of soil and sediment samples by whole-cell sensors and chemical methods yield dissimilar results.

Numerous chemical methods for mimicking the bioavailable fraction of metals have been developed. Most are based on "mild" single or sequential chemical extraction schemes [144], which are applied to release the most easily exchangeable heavy metals. Among the most frequently used extractants is water, which should solubilize the readily mobilizable metals from soil. Stronger extractants, like 1 M KNO_3, $MgCl_2$, or NH_4OAc, or 0.11 M acetic acid, should release the exchangeable metals [143], and are thus likely to generate a fraction of metals that more closely resembles that available to living organism. Although it is assumed that the first extraction step in the sequential extraction procedures yields a bioavailable-mimetic fraction that could be used to predict the hazard of heavy metals to living organisms, few correlational analyses are available to test the validity of this assumption. A comparison of a bioreporter-determined bioavailable fraction and a 0.11 M acetic acid-extracted fraction of Cd and Pb in 60 soil samples collected from the vicinity of metal smelters showed that chemical extraction overestimated the bioavailability of Cd by 23 times, whereas acid-extracted and bioreporter-determined values for Pb were similar [136]. When 1 M NaOAc was used as an extractant, the estimate of Pb bioavailability was four times higher than bioreporter results; in contrast, the bioavailability of Cu was underestimated 10-fold by the same extractant [131]. No clear correlation between the bioavailable and chemically extracted fractions of metals was found in either of these studies. A correlation has been found between $Ca(NO_3)_2$-extractable and bioreporter-measured amounts of Ni, the latter being somewhat lower [80]. Equal fractions of bioreporter-measured Zn and $CaCl_2$-extractable Zn have been reported [145]; however, the sample pool in this study was too small to draw any conclusions on the extent to which calcium salt extracts accurately estimated metal bioavailability.

4.2 *Whole-Cell Bioreporters in Bioavailability Studies*

Nearly every research paper on the subject argues that the ability to assess the bioavailability of heavy metals in environmental matrices is a key motivation for the construction of new microbial reporter strains. The possible environmental applications of whole-cell bioreporters, and their advantages and disadvantages compared to chemical analysis have been thoroughly discussed in a recent review by Harms et al. [146]. Here we present a short overview of bioavailability studies in which bacterial bioreporters were used.

The concept of whole-cell bioreporters as measurement devices to determine bioavailable amounts of metal in environmental samples has been validated for both aqueous and solid (soils and sediments) samples. More novel applications, including determination of As content in rice [147] and Hg content in urine [148] and moss [149] have also been explored using bioreporters.

4.2.1 Water Samples

The most important metals to be tested in water samples are As and Fe. Arsenic testing is important because As is a widespread contaminant of drinking water, and Fe testing is important because Fe is an important nutrient for phytoplankton. Rapid and inexpensive bacterial tests have been developed to test for As in drinking water in developing countries [63, 150–152]. Bioreporter-based Fe bioavailability studies have been conducted on sea and lake water samples to determine the availability of Fe for cyanobacteria [92, 153–155]. Slightly lower As concentrations have been detected in groundwater samples by using bacterial bioreporters than by using chemical methods [151]. However, this is expected as natural waters may contain substances that reduce As bioavailability. For example, silicates and Fe are known to reduce As bioavailability [150, 152]. The reliability of bioreporters in screening drinking water is supported by the observation that only 8% false-negative results were obtained in a set of 194 water samples [152], whereas the false-negative rate with chemical field test kits may be as high as 68% [156].

4.2.2 Solid Samples

Whole-cell bioreporters have been applied to the testing of solid samples, such as soils and sediments. Because a number of processes influence the bioavailability of contaminants in soils and sediments, determining bioavailability in these matrices is a very complicated task, and thus far no universal test has been developed.

Sample preparation is required for soil and sediment samples before bioreporters can be applied. Usually a soil–water mix is prepared since water is essential for the physiological activity of the cell. The addition of water is unavoidable if predried soils are analyzed. Two preparations have typically been used for the analysis of soils and sediments by bacterial bioreporters: particle-containing soil/sediment-water suspensions, and particle-free soil/sediment-water extracts (supernatant from centrifuged soil–water suspensions).

The bioavailability of metals in soil–water extracts can be relatively high; for example, approximately 80% of mercury [106] and up to 100% of As [157, 158] and Cd [120] have been detected in soil–water extracts with bacterial biosensors. In contrast, only 40% of water-extractable Cu [159] and 4–6% of water-soluble Pb in humic surface soil, and 13–43% of water-soluble Pb in mineral soil [133] was available for detection, showing that water–soil extracts may contain significant amounts of metal chelators. However, when bioreporter-detected metal concentrations in soil–water extracts are compared to total metal concentrations in soil, it becomes clear that only a few percent of total metal in soil is bioavailable to bacteria in water extracts [39, 42, 106, 120, 127, 134, 160], indicating that a majority of heavy metals remain adsorbed to soil particles. Sorption of metals to soil particles does not necessarily mean metals are not available to living organisms, as has been demonstrated by soil suspension assays. When bioavailability of metals has been determined directly in soil suspensions, a significantly higher bioavailability of metal in suspension than in particle-free extract (up to 90-fold) has been observed with Hg, Pb, and Cd [39, 120], but not with Zn or Cr [39] or Cu [160]. A higher bioavailability in soil-suspension assays demonstrates the importance of direct contact between the soil matrix and bacteria. Assays employing bacteria immobilized in alginate also showed the importance of direct contact between bacteria and the sample matrix, demonstrating that 20-fold more Hg and 4-fold more As was available to nonimmobilized than alginate-immobilized bioreporter bacteria [15].

In general, bioavailable and total metal concentrations in soils and sediments are not correlated. When samples of the same soil that had been artificially amended to contain different Hg concentrations were tested, there was a good correlation between total and bioavailable Hg [161]. However, when larger sets of different natural soils have been tested, the metal bioavailability varies greatly, from 0.1 to 60% [39, 120, 76].

The large variability in bacterial sensor-determined metal fractions is likely caused to a large extent by physico-chemical properties of the soil, but may also be attributable in part to the particular metal species involved. The considerable adsorption capacity of solid matrices reduces the bioavailability of metals. Adsorption of metal to the soil matrix depends on (1) organic carbon and organic matter content, (2) particle size, (3) pH and redox potential, (4) cation exchange capacity, and (5) temperature and humidity [162]. The sulfide content in sediments also plays an essential role in heavy metal bioavailability [163]. Bioavailability studies of methyl mercury in different soil components showed that five times more Hg was available to bacteria in montmorillonite than in kaolinite or humic acid [164]. Binding of methyl mercury to humic acid is covalent, whereas binding to montmorillonite is predominantly ionic, explaining its greater availability to bacteria [164]. The binding capacities of soil components for other metals probably differ as well. In addition, the porosity of the sample affects the bioavailability of metals. Normally, about 15% of the soil porosity in a sandy soil and 52% in a clayey soil are inaccessible to microorganisms simply due to the small size of pores compared with the bacterial cells [165]. Because the bioavailability of metals in solid matrices is greatly dependent on the properties of a given sample, every case has to be considered individually; no general model can be applied.

4.2.3 Other Applications

Applications of whole-cell bioreporters are not limited to testing contaminated water or soil samples, and determining potential risks of pollution. Whole-cell bioreporters have also been used to monitor changes in bioavailable metal concentrations in different processes, and when used in parallel with other assays are particularly useful in evaluating the efficiency of different remediation/immobilization treatments. For example, a Ni-bioreporter has been helpful in assessing the effects of adding beringite and steel shot to contaminated soil, and the effects of liming on Ni bioavailability [80, 166]. Changes in Zn speciation after remediation with cyclonic ash and compost [145] and changes in Pb bioavailability after treatment with phosphate rock, lime or ash [167] have also been determined using bacterial bioreporters. In addition, the presence of bioavailable Zn and Cu in runoff waters from roofs [132, 168, 169], and release of Cu from antifouling paints into seawater [138] have been studied with whole-cell bioreporters.

Thus, in addition to measuring bioavailability of metals from contaminated sites, the range of bioreporter applications includes the monitoring of different remediation and industrial processes.

4.3 Bioavailability of Heavy Metals to Different Organisms: Generalizability of Bioavailability Data

Bioavailability may differ among different organisms. For example, bacterial cells may take up contaminants through import mechanisms, and sediment-ingesting organisms, such as earthworms, may ingest contaminated soil particles or be exposed to dissolved contaminants by diffusion across the skin. Thus arises the issue of the transferability of bioavailability data from one organism to another [139]. When presenting data or predictions related to contaminant bioavailability, the target organism should always be noted. The use of microorganisms in soil analysis has several advantages. First, they are less costly, requiring very little laboratory space and time. Second, the use of microorganisms in soil analysis is extremely relevant, as microbes serve a number of unique functions in the soil ecosystem (e.g., degrading organic molecules, mobilizing metals).

Few studies have compared heavy metal bioavailability for bacterial bioreporters to that for other organisms in environmental samples and it may be realistic to start by comparing different bioreporters – for example, those based on different host bacteria or different metal-response elements. A comparison of nine different bacterial bioreporters that respond to Hg, Zn, and Cd showed that the bioavailability of these metals to different bacterial species was similar, with one exception: the bioavailability of Hg was higher for Gram-negative than for Gram-positive bacteria [128].

The correlations between results obtained with bacterial bioreporters and eukaryotic bioassays are somewhat more variable. The bioavailability of Cu has proved to be quite similar between bacterial and yeast Cu bioreporters in Cu-polluted soils [131] and in corrosion-induced Cu runoff waters [132]. A linear correlation between the response of a bacterial Ni-biosensor and Ni accumulation in grain and maize plants has also been shown [80]. On the other hand, the bioavailability of Hg to *E. coli* bacterial bioreporters was uncorrelated with its accumulation in plants [161]. However, because the bioreporter results correlated better with total Hg levels, the authors of this latter study concluded that plants are less useful indicators of soil mercury contamination than reporter bacteria.

5 Conclusions/Future Challenges

A considerable number of whole-cell bioreporters are currently available. However, despite the importance of whole-cell bioreporters in determining the biologically relevant fraction of metals in the environment, they are rarely used for actual environmental analyses. Research has tended to focus more on quantity than quality, increasing the number of available bioreporters at the expense of refining bioreporter properties and testing-protocols. Instead of simply searching for new metal-regulated genetic mechanisms that could be used to construct new bioreporters, renewed emphasis should be placed on improving the bioreporters that already exist. The knowledge gained about metal homeostasis in bacteria and metal binding to metal-responsive proteins should be used to improve sensitivity and specificity of metal bioreporters.

To facilitate comparisons between studies, a standardized testing protocol for heavy metal bioreporters should be developed. These efforts should draw attention to the assay conditions, such as bacterial growth phase, growth medium, incubation time, and sample preparation. Of these, the most important consideration is the use of media that do not change the bioavailability of metals in the tested sample. To achieve reliable results from environmental samples, every bioreporter strain should also be accompanied by a noninducible control strain that helps to define the nonspecific effects of the sample on the measured signal. Finally, common data analysis standards should be followed. We suggest the method of Hakkila et al. [53], which presents the signal in a normalized form that shows the fold-induction of the signal above background in the presence of metal, and uses the equation suggested by Long and Winefordner [170] to establish a reliable detection limit that takes into account the variation in the measured signal.

Even the possible adoption of common conditions may not have an immediate impact on the large-scale application of bioreporters to the analysis of metal bioavailability in environmental samples. However, in the long run, these steps are likely to benefit significantly future studies on metal-bioavailability mechanisms and the factors that affect them.

Acknowledgments AH is funded by EnSTe graduate school. MV has been financially supported by Academy of Finland and the Maj and Tor Nessling Foundation.

References

1. Van Dyk TK, Majarian WR, Konstantinov KB, Young RM, Dhurjati PS, LaRossa RA (1994) Rapid and sensitive pollutant detection by induction of heat shock gene-bioluminescence gene fusions. Appl Environ Microbiol 60:1414–1420
2. Belkin S, Smulski DR, Vollmer AC, Van Dyk TK, LaRossa RA (1996) Oxidative stress detection with Escherichia coli harboring a katG'::lux fusion. Appl Environ Microbiol 62: 2252–2256
3. Bechor O, Smulski DR, Van Dyk TK, LaRossa RA, Belkin S (2002) Recombinant microorganisms as environmental biosensors: pollutants detection by Escherichia coli bearing fabA'::lux fusions. J Biotechnol 94:125–132
4. Ptitsyn LR, Horneck G, Komova O, Kozubek S, Krasavin EA, Bonev M, Rettberg P (1997) A biosensor for environmental genotoxin screening based on an SOS lux assay in recombinant Escherichia coli cells. Appl Environ Microbiol 63:4377–4384
5. Vollmer AC, Belkin S, Smulski DR, Van Dyk TK, LaRossa RA (1997) Detection of DNA damage by use of Escherichia coli carrying recA'::lux, uvrA'::lux, or alkA'::lux reporter plasmids. Appl Environ Microbiol 63:2566–2571
6. King JMH, Digrazia PM, Applegate B, Burlage R, Sanseverino J, Dunbar P, Larimer F, Sayler GS (1990) Rapid, sensitive bioluminescent reporter technology for naphthalene exposure and biodegradation. Science 249:778
7. Köhler S, Belkin S, Schmid RD (2000) Reporter gene bioassays in environmental analysis. Fresenius J Anal Chem 366:769–779
8. D'Souza SF (2001) Microbial biosensors. Biosens Bioelectron 16:337
9. Magrisso S, Erel Y, Belkin S (2008) Microbial reporters of metal bioavailability. Microb Biotechnol 1:320–330
10. Meighen EA (1991) Molecular biology of bacterial bioluminescence. Microbiol Rev 55:123–142
11. Viviani VR (2002) The origin, diversity, and structure function relationships of insect luciferases. Cell Mol Life Sci 59:1833
12. Lampinen J, Virta M, Karp M (1995) Comparison of Gram-positive and Gram-negative bacterial strains cloned with different types of luciferase genes in bioluminescence cytotoxicity tests. Environ Toxicol Water Qual 10:41
13. Tsien RY (1998) The green fluorescent protein. Annu Rev Biochem 67:509
14. Tamminen MV, Virta MPJ (2007) Quantification of eco toxicological tests based on bioluminescence using Polaroid film. Chemosphere 66:1329
15. Ivask A, Green T, Polyak B, Mor A, Kahru A, Virta M, Marks R (2007) Fibre-optic bacterial biosensors and their application for the analysis of bioavailable Hg and As in soils and sediments from Aznalcollar mining area in Spain. Biosens Bioelectron 22:1396–1402

16. Daunert S, Barrett G, Feliciano JS, Shetty RS, Shrestha S, Smith-Spencer W (2000) Genetically engineered whole-cell sensing systems: coupling biological recognition with reporter genes. Chem Rev 100:2705–2738
17. Blencowe DK, Morby AP (2003) Zn(II) metabolism in prokaryotes. FEMS Microbiol Rev 27:291–311
18. Nies DH (2003) Efflux-mediated heavy metal resistance in prokaryotes. FEMS Microbiol Rev 27:313–339
19. van der Meer JR, Tropel D, Jaspers M (2004) Illuminating the detection chain of bacterial bioreporters. Environ Microbiol 6:1005–1020
20. Silver S, Phung le T (2005) A bacterial view of the periodic table: genes and proteins for toxic inorganic ions. J Ind Microbiol Biotechnol 32:587–605
21. Frantz B, O'Halloran TV (1990) DNA distortion accompanies transcriptional activation by the metal-responsive gene-regulatory protein MerR. Biochemistry 29:4747–4751
22. Brown NL, Stoyanov JV, Kidd SP, Hobman JL (2003) The MerR family of transcriptional regulators. FEMS Microbiol Rev 27:145–163
23. Tynecka Z, Gos Z, Zajac J (1981) Reduced cadmium transport determined by a resistance plasmid in Staphylococcus aureus. J Bacteriol 147:305–312
24. Laddaga RA, Bessen R, Silver S (1985) Cadmium-resistant mutant of Bacillus subtilis 168 with reduced cadmium transport. J Bacteriol 162:1106–1110
25. Laddaga RA, Silver S (1985) Cadmium uptake in *Escherichia coli* K-12. J Bacteriol 162:1100–1105
26. Makui H, Roig E, Cole ST, Helmann JD, Gros P, Cellier MF (2000) Identification of the Escherichia coli K-12 Nramp orthologue (MntH) as a selective divalent metal ion transporter. Mol Microbiol 35:1065–1078
27. Grass G, Wong MD, Rosen BP, Smith RL, Rensing C (2002) ZupT is a Zn(II) uptake system in Escherichia coli. J Bacteriol 184:864–866
28. Nucifora G, Chu L, Misra TK, Silver S (1989) Cadmium resistance from Staphylococcus aureus plasmid pI258 cadA gene results from a cadmium-efflux ATPase. Proc Natl Acad Sci U S A 86:3544–3548
29. Rensing C, Mitra B, Rosen BP (1997) The zntA gene of Escherichia coli encodes a Zn(II)-translocating P-type ATPase. Proc Natl Acad Sci U S A 94:14326–14331
30. Rensing C, Sun Y, Mitra B, Rosen BP (1998) Pb(II)-translocating P-type ATPases. J Biol Chem 273:32614–32617
31. Lee SW, Glickmann E, Cooksey DA (2001) Chromosomal locus for cadmium resistance in Pseudomonas putida consisting of a cadmium-transporting ATPase and a MerR family response regulator. Appl Environ Microbiol 67:1437–1444
32. Leedjärv A, Ivask A, Virta M (2008) Interplay of different transporters in the mediation of divalent heavy metal resistance in Pseudomonas putida KT2440. J Bacteriol 190:2680–2689
33. Brocklehurst KR, Hobman JL, Lawley B, Blank L, Marshall SJ, Brown NL, Morby AP (1999) ZntR is a Zn(II)-responsive MerR-like transcriptional regulator of zntA in Escherichia coli. Mol Microbiol 31:893–902
34. Binet MR, Poole RK (2000) Cd(II), Pb(II) and Zn(II) ions regulate expression of the metal-transporting P-type ATPase ZntA in Escherichia coli. FEBS Lett 473:67–70
35. Brocklehurst KR, Megit SJ, Morby AP (2003) Characterisation of CadR from Pseudomonas aeruginosa: a Cd(II)-responsive MerR homologue. Biochem Biophys Res Commun 308: 234–239
36. Yoon KP, Misra TK, Silver S (1991) Regulation of the cadA cadmium resistance determinant of Staphylococcus aureus plasmid pI258. J Bacteriol 173:7643–7649
37. Biran I, Babai R, Levcov K, Rishpon J, Ron EZ (2000) Online and in situ monitoring of environmental pollutants: electrochemical biosensing of cadmium. Environ Microbiol 2:285–290
38. Riether K, Dollard M-A, Billard P (2001) Assessment of heavy metal bioavailability using Escherichia coli zntAp::lux and copAp::lux-based biosensors. Appl Microbiol Biotechnol 57:712–716

39. Ivask A, Virta M, Kahru A (2002) Construction and use of specific luminescent recombinant bacterial sensors for the assessment of bioavailable fraction of cadmium, zinc, mercury and chromium in the soil. Soil Biol Biochem 34:1439–1447
40. Tauriainen S, Karp M, Chang W, Virta M (1998) Luminescent bacterial sensor for cadmium and lead. Biosens Bioelectron 13:931–938
41. Shetty RS, Deo SK, Shah P, Sun Y, Rosen BP, Daunert S (2003) Luminescence-based whole-cell-sensing systems for cadmium and lead using genetically engineered bacteria. Anal Bioanal Chem 376:11–17
42. Liao V, Chien M-T, Tseng Y-Y, Ou K-L (2006) Assessment of heavy metal bioavailability in contaminated sediments and soils using green fluorescent protein-based bacterial biosensors. Environ Pollut 142:17–23
43. Rensing C, Grass G (2003) Escherichia coli mechanisms of copper homeostasis in a changing environment. FEMS Microbiol Rev 27:197–213
44. Solioz M, Stoyanov JV (2003) Copper homeostasis in Enterococcus hirae. FEMS Microbiol Rev 27:183–195
45. Solioz M, Odermatt A (1995) Copper and silver transport by CopB-ATPase in membrane vesicles of Enterococcus hirae. J Biol Chem 270:9217–9221
46. Stoyanov JV, Magnani D, Solioz M (2003) Measurement of cytoplasmic copper, silver, and gold with a lux biosensor shows copper and silver, but not gold, efflux by the CopA ATPase of Escherichia coli. FEBS Lett 546:391–394
47. Stoyanov JV, Hobman JL, Brown NL (2001) CueR (YbbI) of Escherichia coli is a MerR family regulator controlling expression of the copper exporter CopA. Mol Microbiol 39:502–511
48. Stoyanov JV, Brown NL (2003) The Escherichia coli copper-responsive copA promoter is activated by gold. J Biol Chem 278:1407–1410
49. Shetty RS, Deo SK, Liu Y, Daunert S (2004) Fluorescence-based sensing system for copper using genetically engineered living yeast cells. Biotechnol Bioeng 88:664–670
50. Dameron CT, George GN, Arnold P, Santhanagopalan V, Winge DR (1993) Distinct metal binding configurations in ACE1. Biochemistry 32:7294–7301
51. Corbisier P, Thiry E, Diels L (1996) Bacterial biosensors for the toxicity assessment of solid wastes. Environ Toxicol Water Qual 11:171–177
52. Tom-Petersen A, Hosbond C, Nybroe O (2001) Identification of copper-induced genes in Pseudomonas fluorescens and use of a reporter strain to monitor bioavailable copper in soil. FEMS Microbiol Ecol 38:59–67
53. Hakkila K, Green T, Leskinen P, Ivask A, Marks R, Virta M (2004) Detection of bioavailable heavy metals in EILATox-Oregon samples using whole-cell luminescent bacterial sensors in suspension or immobilized onto fibre-optic tips. J Appl Toxicol 24:333–342
54. Leskinen P, Virta M, Karp M (2003) One-step measurement of firefly luciferase activity in yeast. Yeast 20:1109–1113
55. Oremland RS, Stolz JF (2003) The ecology of arsenic. Science 300:939–944
56. Rosen BP (2002) Biochemistry of arsenic detoxification. FEBS Lett 529:86–92
57. Ji G, Silver S (1992) Regulation and expression of the arsenic resistance operon from Staphylococcus aureus plasmid pI258. J Bacteriol 174:3684–3694
58. Rosenstein R, Peschel A, Wieland B, Götz F (1992) Expression and regulation of the antimonite, arsenite, and arsenate resistance operon of Staphylococcus xylosus plasmid pSX267. J Bacteriol 174:3676–3683
59. Wu J, Rosen BP (1991) The ArsR protein is a trans-acting regulatory protein. Mol Microbiol 5:1331
60. Corbisier P, Ji G, Nuyts G, Mergeay M, Silver S (1993) Luxab gene fusions with the arsenic and cadmium resistance operons of Staphylococcusaureus plasmid-Pi258. FEMS Microbiol Lett 110:231
61. Tauriainen S, Karp M, Chang W, Virta M (1997) Recombinant luminescent bacteria for measuring bioavailable arsenite and antimonite. Appl Environ Microbiol 63:4456–4461
62. Tauriainen S, Virta M, Chang W, Karp M (1999) Measurement of firefly luciferase reporter gene activity from cells and lysates using Escherichia coli arsenite and mercury sensors. Anal Biochem 272:191–198

63. Stocker J, Balluch D, Gsell M, Harms H, Feliciano J, Daunert S, Malik KA, Van der Meer JR (2003) Development of a set of simple bacterial biosensors for quantitative and rapid measurements of arsenite and arsenate in potable water. Environ Sci Technol 37:4743
64. Ramírez-Díaz MI, Díaz-Pérez C, Vargas E, Riveros-Rosas H, Campos-García J, Cervantes C (2008) Mechanisms of bacterial resistance to chromium compounds. Biometals 21:321–332
65. Branco R, Chung AP, Johnston T, Gurel V, Morais P, Zhitkovich A (2008) The chromate-inducible chrBACF operon from the transposable element TnOtChr confers resistance to chromium(VI) and superoxide. J Bacteriol 190:6996–7003
66. Cervantes C, Ohtake H, Chu L, Misra TK, Silver S (1990) Cloning, nucleotide sequence, and expression of the chromate resistance determinant of Pseudomonas aeruginosa plasmid pUM505. J Bacteriol 172:287–291
67. Alvarez AH, Moreno-Sánchez R, Cervantes C (1999) Chromate efflux by means of the ChrA chromate resistance protein from Pseudomonas aeruginosa. J Bacteriol 181:7398–7400
68. Nies A, Nies DH, Silver S (1990) Nucleotide sequence and expression of a plasmid-encoded chromate resistance determinant from Alcaligenes eutrophus. J Biol Chem 265:5648–5653
69. Peitzsch N, Eberz G, Nies DH (1998) Alcaligenes eutrophus as a bacterial chromate sensor. Appl Environ Microbiol 64:453–458
70. Corbisier P, van der Lelie D, Borremans B, Provoost A, de Lorenzo V, Brown NL, Lloyd JR, Hobman JL, Csoregi E, Johansson G, Mattiasson B (1999) Whole cell- and protein-based biosensors for the detection of bioavailable heavy metals in environmental samples. Anal Chim Acta 387:235–244
71. Navarro C, Wu LF, Mandrand-Berthelot MA (1993) The nik operon of Escherichia coli encodes a periplasmic binding-protein-dependent transport system for nickel. Mol Microbiol 9:1181–1191
72. Eitinger T, Suhr J, Moore L, Smith JA (2005) Secondary transporters for nickel and cobalt ions: theme and variations. Biometals 18:399–405
73. Liesegang H, Lemke K, Siddiqui RA, Schlegel HG (1993) Characterization of the inducible nickel and cobalt resistance determinant cnr from pMOL28 of Alcaligenes eutrophus CH34. J Bacteriol 175:767–778
74. Rodrigue A, Effantin G, Mandrand-Berthelot MA (2005) Identification of rcnA (yohM), a nickel and cobalt resistance gene in Escherichia coli. J Bacteriol 187:2912–2916
75. Schmidt T, Schlegel HG (1994) Combined nickel-cobalt-cadmium resistance encoded by the ncc locus of Alcaligenes xylosoxidans 31A. J Bacteriol 176:7045–7054
76. Tibazarwa C, Wuertz S, Mergeay M, Wyns L, van Der Lelie D (2000) Regulation of the cnr cobalt and nickel resistance determinant of Ralstonia eutropha (Alcaligenes eutrophus) CH34. J Bacteriol 182:1399–1409
77. Rensing C, Pribyl T, Nies DH (1997) New functions for the three subunits of the CzcCBA cation-proton antiporter. J Bacteriol 179:6871–6879
78. García-Domínguez M, Lopez-Maury L, Florencio FJ, Reyes JC (2000) A gene cluster involved in metal homeostasis in the cyanobacterium Synechocystis sp. strain PCC 6803. J Bacteriol 182:1507–1514
79. Grass G, Fan B, Rosen BP, Lemke K, Schlegel HG, Rensing C (2001) NreB from Achromobacter xylosoxidans 31A is a nickel-induced transporter conferring nickel resistance. J Bacteriol 183:2803–2807
80. Tibazarwa C, Corbisier P, Mench M, Bossus A, Solda P, Mergeay M, Wyns L, van der Lelie D (2001) A microbial biosensor to predict bioavailable nickel and its transfer to plants. Environ Pollut 113:19–26
81. Barkay T, Miller SM, Summers AO (2003) Bacterial mercury resistance from atoms to ecosystems. FEMS Microbiol Rev 27:355–384
82. Ralston DM, O'Halloran TV (1990) Ultrasensitivity and heavy-metal selectivity of the allosterically modulated MerR transcription complex. Proc Natl Acad Sci U S A 87:3846–3850
83. Geiselhart L, Osgood M, Holmes DS (1991) Construction and evaluation of a self-luminescent biosensor. Ann N Y Acad Sci 646:53
84. Selifonova O, Burlage R, Barkay T (1993) Bioluminescent sensors for detection of bioavailable Hg(II) in the environment. Appl Environ Microbiol 59:3083–3090

85. Virta M, Lampinen J, Karp M (1995) A luminescence-based mercury biosensor. Anal Chem 67:667–669
86. Yu HR, Mukhopadhyay D, Misra TK (1994) Purification and characterization of a novel organometallic receptor protein regulating the expression of the broad-spectrum mercury-resistant operon of plasmid pDU1358. J Biol Chem 269:15697
87. Ivask A, Hakkila K, Virta M (2001) Detection of organomercurials with whole-cell bacterial sensors. Anal Chem 73:5168–5171
88. Escolar L, Pérez-Martín J, de Lorenzo V (1999) Opening the iron box: transcriptional metalloregulation by the Fur protein. J Bacteriol 181:6223–6229
89. Guerinot ML (1994) Microbial iron transport. Annu Rev Microbiol 48:743–772
90. Bagg A, Neilands JB (1987) Ferric uptake regulation protein acts as a repressor, employing iron (II) as a cofactor to bind the operator of an iron transport operon in Escherichia coli. Biochemistry 26:5471–5477
91. Hantke K (1987) Selection procedure for deregulated iron transport mutants (fur) in Escherichia coli K 12: fur not only affects iron metabolism. Mol Gen Genet 210:135–139
92. Durham KA, Porta D, Twiss MR, McKay RM, Bullerjahn GS (2002) Construction and initial characterization of a luminescent Synechococcus sp. PCC 7942 Fe-dependent bioreporter. FEMS Microbiol Lett 209:215–221
93. Porta D, Bullerjahn GS, Durham KA, Wilhelm SW, Twiss MR, McKay RML (2003) Physiological characterization of a Synechococcus sp. (Cyanophyceae) strain PCC7942 iron-dependent bioreporter. J Phycol 39:64–73
94. Hassler CS, Twiss MR, McKay RML, Bullerjahn GS (2006) Optimization of iron-dependent cyanobacterial (Synechococcus, Cyanophyceae) bioreporters to measure iron bioavailability. J Phycol 42:324–335
95. Joyner DC, Lindow SE (2000) Heterogeneity of iron bioavailability on plants assessed with a whole-cell GFP-based bacterial biosensor. Microbiology 146:2435–2445
96. Mioni CE, Howard AM, Debruyn JM, Bright NG, Twiss MR, Applegate BM, Wilhelm SW (2003) Characterization and field trials of a bioluminescent bacterial reporter of iron bioavailability. Mar Chem 83:31–46
97. Hamlett NV, Landale EC, Davis BH, Summers AO (1992) Roles of the Tn21 merT, merP, and merC gene products in mercury resistance and mercury binding. J Bacteriol 174:6377–6385
98. Pavel H, Forsman M, Shingler V (1994) An aromatic effector specificity mutant of the transcriptional regulator DmpR overcomes the growth constraints of Pseudomonas sp. strain CF600 on para-substituted methylphenols. J Bacteriol 176:7550–7557
99. Garmendia J, Devos D, Valencia A, de Lorenzo V (2001) A la carte transcriptional regulators: unlocking responses of the prokaryotic enhancer-binding protein XylR to non-natural effectors. Mol Microbiol 42:47–59
100. Beggah S, Vogne C, Zenaro E, van der Meer JR (2008) Mutant HbpR transcription activator isolation for 2-chlorobiphenyl via green fluorescent protein-based flow cytometry and cell sorting. Microb Biotechnol 1:68–78
101. DeSilva TM, Veglia G, Porcelli F, Prantner AM, Opella SJ (2002) Selectivity in heavy metal-binding to peptides and proteins. Biopolymers 64:189–197
102. Khan S, Brocklehurst KR, Jones GW, Morby AP (2002) The functional analysis of directed amino-acid alterations in ZntR from Escherichia coli. Biochem Biophys Res Commun 299:438–445
103. Caguiat JJ, Watson AL, Summers AO (1999) Cd(II)-responsive and constitutive mutants implicate a novel domain in MerR. J Bacteriol 181:3462–3471
104. Hughes MN, Poole RK (1991) Metal speciation and microbial growth - the hard (and soft) facts. J Gen Microbiol 137:725–734
105. Rasmussen LD, Turner RR, Barkay T (1997) Cell-density-dependent sensitivity of a mer-lux bioassay. Appl Environ Microbiol 63:3291–3293
106. Fu YJ, Chen WL, Huang QY (2008) Construction of two lux-tagged Hg2+-specific biosensors and their luminescence performance. Appl Microbiol Biotechnol 79:363–370
107. DeLuca M (1978) Methods in enzymology, vol. 57: bioluminescence and chemiluminescence. Academic, New York

108. Collins YE, Stotzky G (1989) Factors affecting the toxicity of heavy metals to microbes. In: Beveridge TJ, Doyle RJ (eds) Metal ions and bacteria. Wiley, New York
109. Gellert G, Stommel A, Trujillano AB (1999) Development of an optimal bacterial medium based on the growth inhibition assay with Vibrio fisheri. Chemosphere 39:467–476
110. Barkay T, Gillman M, Turner RR (1997) Effects of dissolved organic carbon and salinity on bioavailability of mercury. Appl Environ Microbiol 63:4267–4271
111. LaRossa RA, Smulski DR, VanDyk TK (1995) Interaction of lead nitrate and cadmium chloride with Escherichia coli K-12 and Salmonella typhimurium global regulatory mutants. J Ind Microbiol 14:252–258
112. Onishi H, Kobayashi T, Morita N, Baba M (1984) Effect of salt concentration on the cadmium tolerance of a moderately halophilic cadmium-tolerant Pseudomonas sp. Agric Biol Chem 48:2441–2448
113. Farrell RE, Germida JJ, Huang PM (1993) Effects of chemical speciation in growth media on the toxicity of mercury(II). Appl Environ Microbiol 59:1507–1514
114. Nybroe O, Brandt KK, Ibrahim YM, Tom-Petersen A, Holm PE (2008) Differential bioavailability of copper complexes to bioluminescent Pseudomonas fluorescens reporter strains. Environ Toxicol Chem 27:2246–2252
115. Golding GR, Sparling R, Kelly C (2008) Effect of pH on intracellular accumulation of trace concentrations of Hg(II) in Escherichia coli under anaerobic conditions, as measured using a mer-lux bioreporter. Appl Environ Microbiol 74:667–675
116. Kelly CA, Rudd JWM, Holoka MH (2003) Effect of pH on mercury uptake by an aquatic bacterium: implications for Hg cycling. Environ Sci Technol 37:2941–2946
117. Wood KV, DeLuca M (1987) Photographic detection of luminescence in Escherichia coli containing the gene for firefly luciferase. Anal Biochem 161:501–507
118. Petänen T, Virta M, Karp M, Romantschuk M (2001) Construction and use of broad host range mercury and arsenite sensor plasmids in the soil bacterium Pseudomonas fluorescens OS8. Microb Ecol 41:360–368
119. Petänen T, Romantschuk M (2003) Toxicity and bioavailability to bacteria of particle-associated arsenite and mercury. Chemosphere 50:409–413
120. Ivask A, Francois M, Kahru A, Dubourguier HC, Virta M, Douay F (2004) Recombinant luminescent bacterial sensors for the measurement of bioavailability of cadmium and lead in soils polluted by metal smelters. Chemosphere 55:147–156
121. Beveridge TJ, Hughes MN, Lee H, Leung KT, Poole RK, Savvaidis I, Silver S, Trevors JT (1997) Metal-microbe interactions: contemporary approaches. Adv Microb Physiol 38:177–243
122. Hakkila K, Maksimow M, Karp M, Virta M (2002) Reporter genes lucFF, luxCDABE, gfp and dsred have different characteristics in whole-cell bacterial sensors. Anal Biochem 301:235–242
123. Baird GS, Zacharias DA, Tsien RY (2000) Biochemistry, mutagenesis, and oligomerization of DsRed, a red fluorescent protein from coral. Proc Natl Acad Sci U S A 97:11984–11989
124. Bevis BJ, Glick BS (2002) Rapidly maturing variants of the Discosoma red fluorescent protein (DsRed). Nat Biotechnol 20:83
125. Kremers GJ, Goedhart J, van den Heuvel DJ, Gerritsen HC, Gadella TWJ (2007) Improved green and blue fluorescent proteins for expression in bacteria and mammalian cells. Biochemistry 46:3775
126. Leth S, Maltoni S, Simkus R, Mattiasson B, Corbisier P, Klimant I, Wolfbeis OS, Csoregi E (2002) Engineered bacteria based biosensors for monitoring bioavailable heavy metals. Electroanalysis 14:35–42
127. Hansen LJ, Sorensen SJ (2000) Versatile biosensor vectors for detection and quantification of mercury. FEMS Microbiol Lett 193:123–127
128. Bondarenko O, Rõlova T, Kahru A, Ivask A (2008) Bioavailability of Cd, Zn and Hg in soil to nine recombinant luminescent metal sensor bacteria. Sensors 8:6899–6923
129. Harkins M, Porter AJ, Paton GI (2004) The role of host organism, transcriptional switches and reporter mechanisms in the performance of Hg-induced biosensors. J Appl Microbiol 97:1192–1200

130. Park JN, Sohn MJ, Oh DB, Kwon O, Rhee SK, Hur CG, Lee SY, Gellissen G, Kang HA (2007) Identification of the cadmium-inducible Hansenula polymorpha SEO1 gene promoter by transcriptome analysis and its application to whole-cell heavy-metal detection systems. Appl Environ Microbiol 73:5990–6000
131. Peltola P, Ivask A, Åström M, Virta M (2005) Lead and Cu in contaminated urban soils: extraction with chemical reagents and bioluminescent bacteria and yeast. Sci Tot Environ 350:194–203
132. Bahar B, Herting G, Wallinder IO, Hakkila K, Leygraf C, Virta M (2008) The interaction between concrete pavement and corrosion-induced copper runoff from buildings. Environ Monit Assess 140:175–189
133. Turpeinen R, Salminen J, Kairesalo T (2000) Mobility and bioavailability of lead in contaminated boreal forest soil. Environ Sci Technol 34:5152–5156
134. Rasmussen LD, Sorensen SJ, Turner RR, Barkay T (2000) Application of a mer-lux biosensor for estimating bioavailable mercury in soil. Soil Biol Biochem 32:639–646
135. Fritze H, Perkiömäki J, Petänen T, Pennanen T, Romantschuk M, Karp M, Yrjälä K (2001) A microcosms study on the effects of Cd-containing wood ash on the coniferous humus fungal community and the Cd bioavailability. J Soils Sediments 1:146–150
136. Kahru A, Ivask A, Kasemets K, Põllumaa L, Kurvet I, Francois M, Dubourguier HC (2005) Biotests and Biosensors in ecotoxicological risk assessment of field soils polluted with zinc, lead and cadmium. Environ Toxicol Chem 24:2973–2982
137. Heitzer A, Applegate B, Kehrmeyer S, Pinkart H, Webb OF, Phelps TJ, White DC, Sayler GS (1998) Physiological considerations of environmental applications of lux reporter fusions. J Microbiol Methods 33:45–57
138. Sandberg J, Odnevall Wallinder I, Leygraf C, Virta M (2007) Release and chemical speciation of copper from anti-fouling paints with different active copper compounds in artificial seawater. Mater Corros 58:165–172
139. Semple KT, Doick KJ, Jones KC, Burauel P, Craven A, Harms H (2004) Defining bioavailability and bioaccessibility of contaminated soil and sediment is complicated. Environ Sci Technol 38:229A–231A
140. Tauriainen SM, Virta MPJ, Karp MT (2000) Detecting bioavailable toxic metals and metalloids from natural water samples using luminescent sensor bacteria. Water Res 34: 2661–2666
141. Peijnenburg WJGM, Jager T (2003) Monitoring approaches to assess bioaccessibility and bioavailability of metals: matrix issues. Ecotoxicol Environ Saf 56:63–77
142. Adriano DC (2001) Trace elements in terrestrial environments. In: Biogeochemistry, bioavailability, and risks of metals. Springer, Berlin Heidelberg New York
143. Ure AM, Davidson CM (2001) Chemical speciation in soils and related materials by selective chemical extraction. In: Ure AM, Davidson CM (eds) Chemical speciation in environment. Blackwell Science, Oxford
144. Ure AM, Davidson CM, Thomas RP (1995) Single and sequential schemes for trace metal speciation in soil and sediment. In: Quevauviller P, Maier Griepink B (eds) Quality assurance for environmental analysis. Elsevier Science, Amsterdam
145. Nachtegaal M, Marcus MA, Sonke JE, Vangronsveld J, Livi KJT, van der Lelie D, Sparks DL (2005) Effects of in situ remediation on the speciation and bioavailability of zinc in smelter-contaminated soil. Geochim Cosmochim Acta 69:4649–4664
146. Harms H, Wells MC, van der Meer JR (2006) Whole-cell living biosensors – are they ready for environmental application? Appl Microbiol Biotechnol 70:273–280
147. Baumann B, van der Meer JR (2007) Analysis of bioavailable arsenic in rice with whole cell living bioreporter bacteria. J Agric Food Chem 55:2115–2120
148. Roda A, Pasini P, Mirasoli M, Guardigli M, Russo C, Musiani M, Baraldini M (2001) Sensitive determination of urinary mercury(II) by a bioluminescent transgenic bacteria-based biosensor. Anal Lett 34:29–41
149. Pepi M, Reniero D, Baldi B, Barbieri P (2006) A comparison of mer::lux whole cell biosensors and moss, a bioindicator for estimating mercury pollution. Water Air Soil Pollut 173:163–175

150. Harms H, Rime J, Leupin O, Hug SJ, van der Meer JR (2005) Influence of groundwater composition on arsenic detection by bacterial biosensors. Mikrochim Acta 151:217–222
151. Liao VH, Ou KL (2005) Development and testing of a green fluorescent protein-based bacterial biosensor for measuring bioavailable arsenic in contaminated groundwater samples. Environ Toxicol Chem 24:1624–1631
152. Trang PT, Berg M, Viet PH, Van Mui N, van der Meer JR (2005) Bacterial bioassay for rapid and accurate analysis of arsenic in highly variable groundwater samples. Environ Sci Technol 39:7625–7630
153. McKay RML, Porta D, Bullerjahn GS, Al-Rshaidat MMD, Klimowicz JA, Sterner RW, Smutka TM, Brown ET, Sherrell RM (2005) Bioavailable iron in oligotrophic Lake Superior assessed using biological reporters. J Plankton Res 27:1033–1044
154. Hassler CS, Twiss MR (2006) Bioavailability of iron sensed by a phytoplanktonic Fe-bioreporter. Environ Sci Technol 40:2544–2551
155. Boyanapalli R, Bullerjahn GS, Pohl C, Croot PL, Boyd PW, McKay RM (2007) Luminescent whole-cell cyanobacterial bioreporter for measuring Fe availability in diverse marine environments. Appl Environ Microbiol 73:1019–1024
156. Rahman MM, Mukherjee D, Sengupta MK, Chowdhury UK, Lodh D, Chanda CR, Roy S, Selim M, Quamruzzaman Q, Milton AH, Shahidullah SM, Rahman MT, Chakraborti D (2002) Effectiveness and reliability of arsenic field testing kits: are the million dollar screening projects effective or not? Environ Sci Technol 36:5385–5394
157. Petänen T, Lyytikainen M, Lappalainen J, Romantschuk M, Kukkonen JVK (2003) Assessing sediment toxicity and arsenite concentration with bacterial and traditional methods. Environ Pollut 122:407–415
158. Turpeinen R, Virta M, Häggblom MM (2003) Analysis of arsenic bioavailability in contaminated soils. Environ Toxicol Chem 22:1–6
159. Brandt KK, Petersen A, Holm PE, Nybroe O (2006) Decreased abundance and diversity of culturable Pseudomonas spp. populations with increasing copper exposure in the sugar beet rhizosphere. FEMS Microbiol Ecol 56:281–291
160. Brandt KK, Holm PE, Nybroe O (2006) Bioavailability and toxicity of particle-associated copper as determined by two bioluminescent Pseudomonas fluorescens biosensor strains. Environ Toxicol Chem 25:1738–1741
161. Bontidean I, Mortari A, Leth S, Brown NL, Karlson U, Larsen MM, Vangronsveld J, Corbisier P, Csoregi E (2004) Biosensors for detection of mercury in contaminated soils. Environ Pollut 131: 255–262
162. Tarradellas J, Bitton G, Rossel D (1997) Soil ecotoxicology. Lewis, Florida
163. Di Toro DM, Mahony JD, Hanson DJ, Scott KJ, Hicks MB, Redmond MS (1990) Toxicity of cadmium in sediments: the role of acid volatile sulfide. Environ Toxicol Chem 9: 1489–1504
164. Bernaus A, Gaona X, Ivask A, Kahru A, Valiente M (2005) Analysis of sorption and bioavailability of different species of mercury on model soil components using XAS techniques and sensor bacteria. Anal Bioanal Chem 382:1541–1548
165. Huang PM, Bollag J-M, Senesi N (2002) Interactions between soil particles and microorganisms. Wiley, New York
166. Everhart JL, McNear D Jr, Peltier E, van der Lelie D, Chaney RL, Sparks DL (2006) Assessing nickel bioavailability in smelter-contaminated soils. Sci Total Environ 367:732–744
167. Geebelen W, Adriano DC, van der Lelie D, Mench M, Carleer R, Clijsters H, Vangronsveld J (2003) Selected bioavailability assays to test the efficacy of amendment-induced immobilization of lead in soils. Plant Soil 249:217–228
168. Heijerick DG, Janssen CR, Karlèn C, Wallinder IO, Leygraf C (2002) Bioavailability of zinc in runoff water from roofing materials. Chemosphere 47:1073–1080
169. Karlen C, Wallinder IO, Heijerick D, Leygraf C (2002) Runoff rates, chemical speciation and bioavailability of copper released from naturally patinated copper. Environ Pollut 120:691–700
170. Long GL, Winefordner JD (1983) Limit of detection: a closer look at the IUPAC definition. Anal Chem 55:712A–724A

Adv Biochem Engin/Biotechnol (2010) 118: 65–84
DOI: 10.1007/10_2009_7

Published online: 21 May 2009

Bacteriophage-Based Pathogen Detection

Steven Ripp

Abstract Considered the most abundant organism on Earth, at a population approaching 10^{31}, bacteriophage, or phage for short, mediate interactions with myriad bacterial hosts that has for decades been exploited in phage typing schemes for signature identification of clinical, food-borne, and water-borne pathogens. With over 5,000 phage being morphologically characterized and grouped as to susceptible host, there exists an enormous cache of bacterial-specific sensors that has more recently been incorporated into novel bio-recognition assays with heightened sensitivity, specificity, and speed. These assays take many forms, ranging from straightforward visualization of labeled phage as they attach to their specific bacterial hosts to reporter phage that genetically deposit trackable signals within their bacterial hosts to the detection of progeny phage or other uniquely identifiable elements released from infected host cells. A comprehensive review of these and other phage-based detection assays, as directed towards the detection and monitoring of bacterial pathogens, will be provided in this chapter.

Keywords Bacteriophage • Pathogen • Phage • Phage amplification • Phage display • Reporter gene

Contents

S. Ripp
The University of Tennessee, 676 Dabney Hall, Knoxville, TN 37996, USA
e-mail: saripp@utk.edu

1 The Bacteriophage Lifestyle

Prior to discussing ways in which bacteriophage can be used to monitor and detect bacterial pathogens, it is first necessary to understand how phage recognize, infect, and propagate within their bacterial hosts. Phage possess a host range that represents the types of bacterial cells they can infect. Host ranges can be diverse, infecting across bacterial strains, species, and genera, or highly specific, infecting only within a single bacterial serotype. This broadness or narrowness of bacterial host recognition is dependent on bacterial surface receptors that the phage uniquely identifies, and, depending on the phage, can consist of surface structures such as flagella or pili, surface polysaccharides, or a diverse range of surface or membrane attached proteins (Fig. 1). By knowing a phage's host range, one can identify bacteria based on phage infection patterns, in what is referred to as phage typing. It is the fundamentals of phage typing upon which phage-mediated pathogen detection and

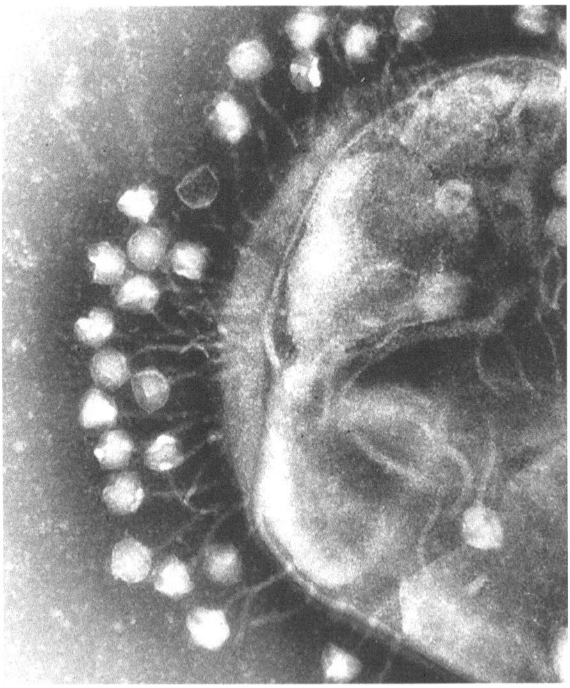

Fig. 1 Electron micrograph of bacteriophage attaching to a host bacterial cell. From Wikipedia.com

recognition is based. After spreading a culture of an unknown bacterium on a solid media petri plate, small drops of different phage solutions are added, and resulting productive infections are identified by the formation of a zone of clearing, or a plaque, representing areas where bacterial hosts were killed, or lysed, by the phage. Based on which phage infect and form plaques and which phage do not, the unknown bacterium can be epidemiologically identified. Phage typing schemes are widely available for most pathogenic microorganisms and their use in diagnostic microbial identification down to serotype or serogroup classifications has been well documented. This includes phage typing sets for virtually all National Institute of Allergy and Infectious Diseases (NIAID) Category A, B, and C microbial pathogens, including *Yersinia enterocolitica*, *Y. pestis*, *Bacillus anthracis*, *Campylobacter jejuni*, diarrheagenic *Escherichia coli*, *Listeria monocytogenes*, *Staphylococcus*, *Clostridium botulinum*, *C. perfringens*, *Vibrio cholerae*, *Salmonella*, *Mycobacterium tuberculosis*, *Brucella* sp., *Burkholderia* sp., and *Shigella* sp.

Once the phage recognizes an appropriate host bacterial cell, it infects that cell via insertion of its nucleic acid and instigates either a lytic or lysogenic interaction. A phage labeled as virulent (or lytic, or productive) will undergo a lytic interaction with its host. A phage labeled as temperate (or reductive) will undergo a lysogenic interaction. In the lytic response, the phage's nucleic acid takes over the host cell, committing it towards the production and assembly of new progeny phage particles that are subsequently released when the host cell bursts open upon lysis. These progeny phage then continue to infect and propagate themselves within new bacterial hosts. In lysogeny, the phage nucleic acid coexists and replicates with its bacterial host as a prophage, and at a later time can enter the lytic cycle in order to propagate itself. Some phage-mediated detection assays rely on lytic phage, others rely on temperate phage, and for others either will suffice. It is important to note that no matter what the phage, the host cell must be alive for infection and propagation to occur. Therefore, unlike nucleic acid-based (i.e., the polymerase chain reaction (PCR)) or immunological-based (i.e., enzyme-linked immunosorbent assay (ELISA)) pathogen detection schemes where living targets cannot be distinguished from dead targets, mainstream phage-mediated detection assays identify only living bacterial targets.

2 Bacteriophage-Mediated Pathogen Detection Schemes

With the understanding of a phage's host range and with decades of historical experience in the application of phage typing schemes, it was evident that phage could successfully function as bacterial-specific indicators. Consequently, phage-mediated detection assays more efficient, sensitive, and faster than the venerable phage plaque assays were developed. These detection schemes exploit all facets of the phage/host interaction, with some relying on the initial recognition and attachment of the phage to the host cell, while others depend on the infective insertion of phage nucleic acid into the host cell, or genetic expression of the phage nucleic acid, or the actual propagation and release of progeny phage upon host cell lysis. These varied approaches, however, are all functionally dependent on and do not deviate from the fundamentals of phage/host recognition and the host range characteristics unique to each phage.

2.1 Reporter Phage

Reporter phage carry with them an easily visualized signal that indicates when a productive infection has occurred. By choosing a phage specific for the desired bacterial target, the presence or absence of a signal denotes the parallel presence or absence of the bacterial target. Signals are endowed to the phage through the use of reporter genes whose sole purpose is to generate a reporter protein that yields an easily discernable reporter signal, most often of the bioluminescent, fluorescent, or chemiluminescent variety, although other more unusual signals such as ice formation have been incorporated as well. The reporter gene is placed directly within the phage's genome, where, upon infection, it is transferred to the host cell and subsequently expressed to yield the signal.

2.1.1 Green Fluorescent Protein Reporter Phage

Reporter phage carrying the green fluorescent protein (*gfp*) reporter gene emit a 508 nm fluorescent signal when activated by ultraviolet or blue light. Funatsu et al. [1] first demonstrated *gfp* signaling in phage lambda for the general detection of *E. coli*. Reporter phage with the *gfp* gene inserted within their genome were combined with a mixed culture of *E. coli* and *Mycobacterium smegmatis* for 4 h and then observed under an epifluorescent microscope. *E. coli* cells infected with the phage and thus now containing the *gfp* gene fluoresced while noninfected *M. smegmatis* cells did not, demonstrating that this reporter assay could distinctly identify target *E. coli* cells within a mixed culture environment.

Tanji et al. [2] constructed another *E. coli* specific *gfp* reporter phage using phage T4. Wild-type phage T4 is lytic, and therefore destroys the *E. coli* cells it infects. This is advantageous in some respects because the pathogenic target can not only be detected but additionally killed. However, sensitivity suffers since the assay relies on the detection of phage-derived fluorescence from *E. coli* cells, and destroying the targets one wishes to detect is not effective. They therefore inactivated the lytic activity of phage T4 by eliminating its ability to synthesize the lysozyme enzyme responsible for host cell wall degradation that leads to cell lysis. This reporter phage, designated T4e$^-$/GFP, when added to a mixed culture of *E. coli* cells and *Pseudomonas aeruginosa* cells, selectively identified only the target *E. coli* cells within a 1-h assay (Fig. 2). Miyanaga et al. [3] later used these reporter phage to detect *E. coli* directly in sewage influent. Since the host range of reporter phage T4e$^-$/GFP was not inclusive of all *E. coli* present in the sewage

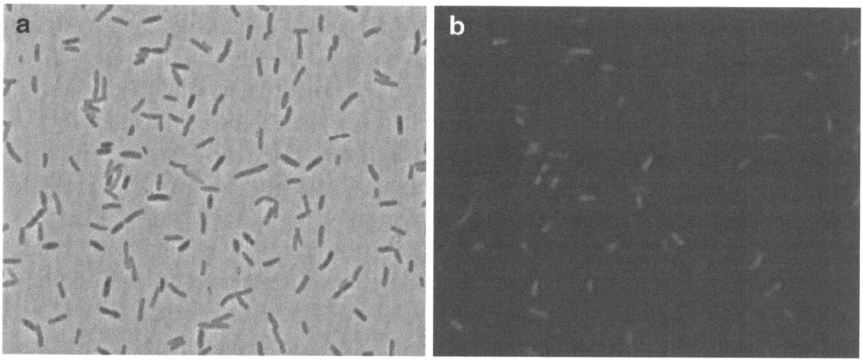

Fig. 2 **a** Optical micrograph of a mix of *E. coli* and *P. aeruginosa* cells. **b** A same field-of-view fluorescent microscopic image demonstrating fluorescent identification of *E. coli* cells after addition of the *E. coli* specific T4e$^-$/GFP reporter phage. Used with permission from Tanji et al. [2]

samples, only a small percentage (~8%) of the total *E. coli* population was detected, which was not a surprising outcome since the use of a single reporter phage severely confines target acquisitions. To encompass all desired targets, it is probably necessary with most reporter phage assays to incorporate several reporter phage whose cumulative host ranges sufficiently converge across the suite of bacterial targets requiring detection. Namura et al. [4] went on to do this by isolating two other phage from sewage and genetically incorporating each with *gfp* reporter genes. Together, these reporters demonstrated a host range covering nearly 50% of the *E. coli* sewage isolates.

Oda et al. [5] created a much narrower host range *gfp* reporter phage using the *E. coli* O157:H7-specific phage PP01 which was capable of discriminating between *E. coli* O157:H7 and *E. coli* K12 cells within 10 min based on fluorescence emission. Sensitivity was later improved by once again inactivating the lytic activity of the phage [6], providing clearer and sharper epifluorescent images that translated into easier and more direct identification of *E. coli* O157:H7 in mixed cultures. These assays additionally discriminated between healthy and stressed cells, where healthy cells emitted bright green fluorescence while metabolically stressed cells emitted faded fluorescent signals. This allowed for easy, simultaneous identification of healthy cells vs cells within a viable but nonculturable (VBNC) state which, when using conventional plating methods, cannot be accomplished without supplemental and time-consuming steps.

2.1.2 Lux Reporter Phage

The *lux* genetic system, based on the bacterial luciferase derived most typically from the microbe *Vibrio fischeri*, generates a 490-nm bioluminescent light signal that can be detected with any variety of photomultiplier tube (PMT) or charge-coupled device (CCD)-based instruments [7]. Most reporter phage assays incorporate two of the *lux* genes, *luxA* and *luxB* (or, more simply, *luxAB*) within the phage genome, and the passage of these genes to the target bacterial host in combination with an added aldehyde (*n*-decanal) substrate, generates an immediate burst of bioluminescent light. Ulitzur and Kuhn [8] first demonstrated *lux*-based reporter phage detection using a *luxAB* incorporated phage λ Charon 30 to detect *E. coli*. The reporter phage itself does not possess the metabolic machinery necessary to express the *luxAB* genes, and therefore remains "dark." Only after infection and insertion of the *luxAB* genes into the metabolically active host is there transcriptional and translational activation and subsequent bioluminescent output, contingent on *n*-decanal addition. With this *luxAB* reporter phage, they were able to detect between 10 and 100 *E. coli* cells per milliliter in artificially contaminated milk or urine within 1 h. Under more real-world conditions, using swab samples from slaughterhouse surfaces and swine carcasses, Kodikara et al. [9] demonstrated detection limits of 10^4 *E. coli* cells cm^{-2} or g with this reporter phage assay. Detection limits were further improved with a 4-h preenrichment, permitting detection down to 10 cells cm^{-2} or g.

Other *luxAB* reporter phage have since been engineered for the detection of additional pathogenic microorganisms. Waddell and Poppe [10] inserted the *luxAB* genes into phage ΦV10 for the specific detection of *E. coli* O157:H7 within 1 h. Chen and Griffiths [11] constructed several P22-based *luxAB* reporter phage specific for the A, B, and D_1 *Salmonella enterica* serotypes. *Salmonella* isolates in pure culture were detected down to 10 cfu mL^{-1} after a 6-h preenrichment. Thouand et al. [12] further evolved the assay to a more commercially viable kit format with testing in various poultry feed, feces, and litter samples artificially inoculated with *Salmonella typhimurium* (*S. enterica* subspecies *enterica* serovar Typhimurium). Within the approximate 16 h assay format, bacterial concentrations enriched above 10^6 cfu mL^{-1} could be detected via bioluminescent light emission. Although not a rapid assay due to the preenrichment step, it does surpass the normal 24 h or greater incubation times required by standard selective plating methods and would ultimately be a very straightforward hands-off assay if adapted to robotic liquid handling. In another application of this reporter phage, Chen and Griffiths [11] injected *Salmonella enteritidis* directly into poultry eggs at a lower inoculum of 63 cfu mL^{-1}. After a 24-h incubation and *n*-decanal addition, a BIQ Bioview Image Quantifier camera system was used to image the eggs directly. Phage infected bioluminescent *Salmonella* cells could be in situ visualized through the unadulterated egg shell and localized as to their area of infection within the whole egg.

A *luxAB* reporter phage for *Listeria monocytogenes* was created by Loessner et al. [13] using phage A511 which infects approximately 95% of the serovars responsible for human listeriosis. *L. monocytogenes* at concentrations between 0.1 and 1,000 cfu g^{-1} was artificially inoculated into several food sources such as cheeses, chocolate pudding, cabbage, lettuce, ground beef, liverwurst, milk, and shrimp. After a 20-h preenrichment, the A511 *luxAB* reporter phage and *n*-decanal substrate were added to yield detection limits as low as one cell g^{-1}. Foods with more complex microbial background flora, such as ground beef, yielded detection limits of 10 cells g^{-1}. A total of 348 naturally contaminated meats, poultry, dairy products, and other environmental samples were also assayed in parallel with standard plating techniques, and *Listeria* positive samples were found to correspond between the two methods. With standard plating requiring 72-96 h, the less than 24 h A511 *luxAB* reporter phage assay was shown to be highly efficient while generating comparable results.

Ripp et al. [14] used *lux* in a uniquely different manner in phage reporter assays designed to detect *E. coli*. Conventional *luxAB* phage reporters require the addition of *n*-decanal before bioluminescence can be generated. This can be disadvantageous since it permits only single time point rather than continuous measurements. Besides the *luxAB* genes, the *lux* genetic system also includes the *luxC*, *luxD*, and *luxE* genes which in the wild-type *V. fischeri* cell are responsible for generating the *n*-decanal aldehyde substrate that in concert with cellular $FMNH_2$ produces self-generated bioluminescent light. Reporter systems that use all of the *lux* genes (*luxCDABE*) therefore do not require *n*-decanal addition and can emit bioluminescent light continuously once activated. Due to the size of the *luxCDABE* gene cassette and the limits as to how much nucleic acid a phage head can accommodate, its insertion into a phage has yet to be achieved. However, the *luxCDABE* genes can be inserted into bacterial cells to form living bacterial bioreporters that sense and respond to chemical targets, usually directed towards environmental monitoring and the detection of contaminants such as heavy metals, pesticides, polycyclic aromatic hydrocarbons (PAHs), etc. [15]. The Ripp group first constructed a *luxCDABE*-based bacterial bioluminescent bioreporter, referred to as OHHLux, that bioluminesced in the presence of the quorum sensing chemical *N*-3-(oxohexanoyl)-l-homoserine lactone (OHHL). OHHL is generated by another *lux* gene, *luxI*, which was inserted into the genome of phage lambda. When this lambda-*luxI* reporter phage infects *E. coli*, the host *E. coli* cell expresses the *luxI* gene and begins synthesizing OHHL. OHHL diffuses out of the *E. coli* cell into the surrounding medium where it interacts with the OHHLux bioluminescent bioreporter and triggers it to autonomously generate bioluminescence (Fig. 3a). Thus, the bioluminescent signal is linked to and indicates the initial phage/host infection event. In pure culture experiments, the assay detected *E. coli* at 1 cfu mL^{-1} within

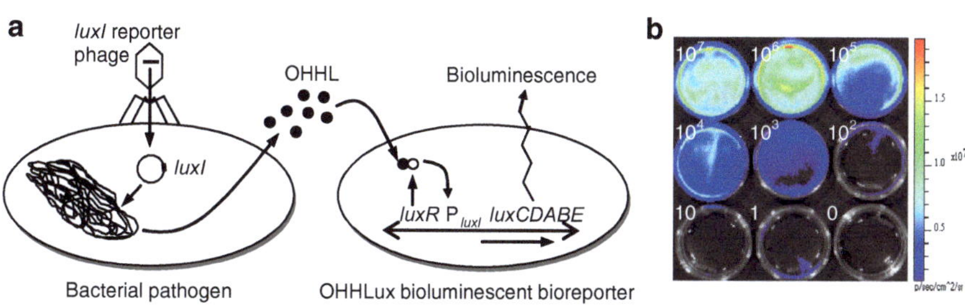

Fig. 3 **a** By incorporating the quorum sensing related *luxI* gene into the reporter phage, infected bacterial pathogens are directed towards synthesis of signature autoinducer chemical molecules (i.e., OHHL) that can be autonomously detected by a *luxCDABE*-based bacterial bioluminescent bioreporter. **b** The combination of *luxI* reporter phage with bioluminescent bioreporters and *E. coli* O157:H7 target host cells permits real-time in situ imaging of bioluminescence emission, shown here in spinach leaf rinsates artificially contaminated from 10^7 to 1 cfu mL^{-1} after a 6-h exposure

approximately 10 h. In washings from artificially contaminated leaf lettuce, a detection limit of approximately 100 cfu mL^{-1} was achieved in 22 h. Since no user manipulation is required, besides combining the phage reporter and bioluminescent bioreporter with the sample, these assays are decidedly user-friendly. The assay was later defined to the specific detection of *E. coli* O157:H7 through similar *luxI* insertion into phage PP01 and tested in artificially contaminated apple juice (detection limit of 1 cfu mL^{-1} within 22 h) and tap water (detection limit of 1 cfu mL^{-1} within 12.5 h) [16]. Additionally, bioluminescent imaging was used to monitor continuously light emission from artificially contaminated spinach leaf rinsate. After a 2-h preenrichment, *E. coli* O157:H7 was separated from the natural microbial biomass with immunomagnetic separation using paramagnetic beads coated with polyclonal antibodies against *E. coli* O157:H7 (Dynal Dynabeads™). The resulting samples were then imaged in situ in real-time under a Xenogen IVIS Lumina camera, resulting in detection down to 1 cfu mL^{-1} within a total assay time of 6 h (Fig. 3b).

2.1.3 Luc Reporter Phage

The *luc* genes encode for the eukaryotic luciferase that catalyzes the two-step conversion of D-luciferin to oxyluciferin to generate a 560 nm bioluminescent light signal. Analogous to the addition of *n*-decanal in *luxAB* reporter assays, the d-luciferin substrate must also be added exogenously during *luc*-based reporter assays. The *luc* genes were first isolated from the firefly *Photinus pyralis*, and, due to their similarities to *lux*, are sometimes referred to as *fflux*. The use of *luc* in reporter phage has primarily been associated with *Mycobacterium* specific phages and the detection and assessment of drug susceptibility in *M. tuberculosis* [17, 18]. Susceptibility to antimycobacterial drugs can be determined by comparing bioluminescent output kinetics from reporter phage added to antibiotic-free or antibiotic-amended *Mycobacterium* cultures. If the drug is effective, fewer host *Mycobacterium* cells are available for phage infection and less light is therefore emitted as compared to the antibiotic-free control. These assays can require several days to complete, but far surpass the several week time periods required using conventional antibiotic susceptibility assays. Bardarov et al. [19] used *luc* reporter phage phAE142 to assay sputum samples for *M. tuberculosis* presence and establish antibiotic susceptibility profiles. *M. tuberculosis* could be detected within a median of 7 days at concentrations greater than 10^4 cfu mL^{-1}. Standard mycobacterial growth indicator tube (MGIT) assays performed better at lower concentrations but with detection occurring within a median of 9 days. For antibiotic susceptibility testing, results could be obtained in 3 days using reporter phage while 12 days were required using standard tests. Riska et al. [20], recognizing the simplicity of these phage reporter assays and the prevalence of tuberculosis in developing and under-developed countries, developed a low-cost Polaroid film-based device, referred to as the Bronx Box, for recording bioluminescence emission kinetics. The device combines photographic film with a multiwell microtiter sample plate. Positive wells emitting bioluminescent light (i.e., those containing *luc* reporter phage and viable, infectable *M. tuberculosis* cells that had not been affected by the antibiotic) can be identified with corresponding spots of exposure on the Polaroid film. Using clinical isolates, there was a 100% correspondence in antibiotic susceptibility profiles between the Bronx Box and standard laboratory methods, with the Bronx Box providing results in 94 h as compared to 3 weeks for the standard methods [21].

2.1.4 LacZ Reporter Phage

The *lacZ* gene encodes a β-galactosidase enzyme that catalyzes the hydrolysis of β-galactosides. With the addition of a fluorescent, luminescent, or chemiluminescent substrate, the reaction endpoint can be followed in a typical reporter gene fashion. Goodridge and Griffiths [22] successfully inserted the *lacZ* reporter gene into phage T4 for biosensing of *E. coli*. After addition of a chemiluminescent substrate to phage infected *E. coli* cells, they were able to achieve detection down to 100 cfu mL^{-1} in pure culture within 12 h. The *lacZ* reporter gene has, however, seen very little additional use in reporter phage assays, with better performing *gfp*, *lux*, and *luc* assays taking precedent.

2.1.5 InaW Ice Nucleation Reporter Phage

The *inaW* gene encodes for ice nucleation. Its genetic expression yields the InaW protein which integrates itself into the outer cellular membrane where it catalyzes ice crystal formation at temperatures between −2 and −10 °C. Therefore, *inaW* functions as a reporter gene whose end product signal is ice formation at supercooled temperatures. Wolber and Green [23] inserted the *inaW* gene into *Salmonella* phage P22 to create the BIND™ (bacterial ice nucleation diagnostic) assay. When this reporter phage infects *Salmonella*, the *inaW* gene is expressed and resulting assay samples, when cooled to −9.5 °C, freeze. In the BIND™ assay, an indicator dye turns orange if freezing occurs and fluorescent green if it does not, thereby providing a simple colorimetric or fluorescent signaling endpoint. When tested in artificially contaminated foods such as raw egg and milk, detection limits of less than 10 cells mL^{-1} were obtained. The BIND™ assay was sold as a commercial kit but is no longer marketed.

2.2 *Labeled Phage*

Rather than requiring the phage to infect its host cell, labeled phage simply rely on the initial attachment of the phage to its host cell to signal host cell presence. Goodridge et al. [24, 25] demonstrated this technique by staining the DNA of the *E. coli* O157:H7 specific phage LG1 with the fluorescent dye YOYO-1. Upon attachment of this fluorescently labeled phage to a suitable *E. coli* O157:H7 host, a fluorescent halo could be visualized around the cell when viewed with an epifluorescent microscope. When confronted with a mixed bacterial culture, immunomagnetic separation was first used to separate target *E. coli* cells from the mix, and then fluorescently labeled LG1 phage was added. Flow cytometry permitted rapid enumeration of phage attached *E. coli* cells. Detection limits in artificially contaminated ground beef and raw milk approached approximately 2 cfu g^{-1} after a 6-h enrichment and 10 cfu mL^{-1} after a 10-h enrichment, respectively. Kenzaka et al. [26] demonstrated a similar procedure using DAPI (4′,6-diamidino-2-phenylindole) labeled phage T4 to detect *E. coli*. Using epifluorescence microscopy, they successfully enumerated *E. coli* within 30 min in water samples obtained from a fecally contaminated canal in Thailand. Mosier-Boss et al. [27] labeled P22 phage with fluorescent SYBR gold dye. In this case, the labeled DNA was visualized after phage-mediated injection into target *S. typhimurium* host cells.

2.3 Phage Amplification

Phage amplification assays exploit the ultimate outcome of phage infection, that is, cell lysis and the release of progeny phage, to signal the presence of a specific bacterial pathogen. In its most basic application, wild-type lytic phage are added to a bacterial culture. If suitable host cells are present, infection occurs, phage reproduce, and host cells lyse to release amplified numbers of progeny phage. This burst of new phage synthesis or "phage amplification" therefore indicates that host cells compatible with the phage's host range were present. Hirsh and Martin [28, 29] first demonstrated this technique in the early 1980s using phage Felix-01 to detect *Salmonella*. If the sample contained *Salmonella* cells, then resulting phage infection generated numerous progeny phage that could be detected by a unique signature peak after high-performance liquid chromatography (HPLC) analysis. With poor detection limits (10^6 *Salmonella* g^{-1} or mL^{-1} sample) and the complexities associated with HPLC, these phage amplification assays never gained favor. However, over a decade later, Stewart et al. [30] demonstrated a much simpler and more sensitive phage amplification assay again using phage Felix O-1 for detecting *Salmonella* as well as phages NCIMB 10116 and 10884 for detecting *P. aeruginosa*. These phage were combined with bacterial cultures and incubated for approximately 25 min to allow for phage/host infection to materialize. Free phage remaining in solution were then destroyed via addition of a pomegranate rind virucide. Several minutes later, the virucide was neutralized and "helper cells" were added to provide fresh infectable hosts for progeny phage newly released from infected cells, which were then enumerated by plaque counts using a standard top agar plating method. Within 4 h, as few as 600 *S. typhimurium* or 40 *P. aeruginosa* cells per milliliter could be detected in pure culture experiments. The significant advantage inherent in phage amplification assays is that almost any phage can be used, so long as it is lytic, and the chosen phage remains in its wild-type state, so no genetic engineering or other manipulations are required.

Phage amplification assays have realized their widest application with the commercially available *FASTPlaque* TB™ kit for the clinical detection of *Mycobacterium tuberculosis* (Fig. 4) [31]. The phage, referred to as Actiphage™, is added to the sample for 1 h followed by the addition of the virucide. Five minutes later the virucide is neutralized and a rapid growing mycobacterial cell suspension is added (Sensor™ cells) to promote additional phage infection and amplification. Resulting phage are enumerated with plaque counts on top agar plates. Two large-scale studies verified the detection of 65-83% of confirmed *M. tuberculosis* infections in sputum samples within 2 days using this assay [32]. A *FASTPlaque-Response*™ kit is also available for establishing rifampicin resistant *M. tuberculosis*. The sample is preincubated in the presence or absence of rifampicin antibiotic and then subjected to the phage amplification assay. If the cells are resistant to rifampicin, the number of plaques enumerated will be similar in both samples. If the cells are sensitive to rifampicin, the number of plaques in the rifampicin treated sample will be less than the rifampicin free sample. Susceptibility testing for various other antituberculosis drugs (isoniazid, ethambutol, streptomycin, pyrazinamide, ciprofloxacin) can additionally be achieved [31].

One drawback of phage amplification is that there is no universally effective virucide. Consequently, each phage/host combination must be pretested with any number of virucidal agents to ensure that the virucide inactivates free phage and not host bacterial cells. Favrin et al. [33] circumvented the use of virucides altogether in their assays using phage SJ2 for the detection of *S. enteritidis*. *Salmonella* cells

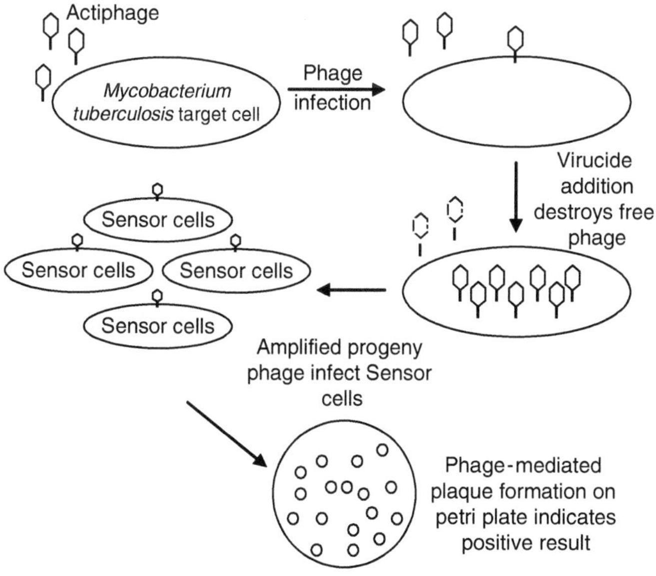

Fig. 4 The *FastPlaque* TB™ test uses phage infection and subsequent amplification of progeny phage to indicate the presence of target *Mycobacterium tuberculosis* host cells

were first concentrated using immunomagnetic separation and then incubated with phage SJ2 for 10 min. With the paramagnetic beads capturing and sequestering the cells, further washing steps removed only free phage. Remaining cells, if phage infected, then released new progeny phage that then reinfected freshly added *Salmonella* "signal-amplifying cells" (SACS) to yield amplified numbers of phage. Rather than enumerating phage by plaque counts, a faster technique based on optical density was used where a decrease in optical density indicated that signal amplifying cell concentrations were declining due to infection and lysing by phage while an increase in optical density indicated an unaffected and growing population of signal-amplifying cells. The assay was performed in artificially contaminated skimmed milk powder, chicken rinses, and ground beef, with an average detection limit of 3 cfu g^{-1} or mL within a total assay time of 20 h, inclusive of preenrichment incubations [34]. It was also applied towards the detection of *E. coli* O157:H7 using phage LG1 and anti *E. coli* paramagnetic beads, with a detection limit of 2 cfu g^{-1} ground beef within an assay time of 23 h [34].

The disadvantage of paramagnetic beads is, however, their expense, and routine use can be cost prohibitive. Jassim and Griffiths [35] rather relied on fluorochromic stains to detect target host cell viability associated with phage infection and lysis. The *Bac* Light live/dead assay from Molecular Probes uses SYTO9 dye to stain viable cells fluorescent green and propidium iodide to stain dead cells fluorescent red. Since propidium iodide penetrates only those cells with damaged cell membranes, as would occur after phage infection, phage infected cells can be discriminated based on color. After performing the phage amplification assay, each dye is added to the sample and the resulting ratio of green to red fluorescent cells indicates the presence of bacterial cells susceptible to phage infection. Phage NCIMB 10116 was used in such a fashion to detect *P. aeruginosa* targets down to 10 cfu mL^{-1} in pure culture within 4 h with no preenrichment.

Ulitzur and Ulitzur [36] avoided altogether the removal of noninfecting phage in their assays by using novel phage mutant repair mechanisms to ensure that endpoint plaque formation was due only to infected target bacteria. Phage carrying amber mutations (phage Felix-O1 for *Salmonella*), ultraviolet light irradiated mutations (phage OE for *E. coli*), or temperature sensitive mutations (phage AR1 for *E. coli* O157:H7) were genetically constructed. These phage could not form plaques on their host cells unless their mutations were repaired by recombination or complementation, thereby negating the need to wash and/or centrifuge the assay samples to remove free phage. For example, two temperature sensitive phage mutants were allowed to coinfect their *E. coli* O157:H7 targets at the permissive temperature (37 °C), and then incubated at their restrictive temperature (42 °C) to prevent further infection cycles. Subsequent plaque formation was therefore only possible if the mutation had been repaired since any remaining mutant phage, due to their temperature sensitivity, could not plaque at 42 °C, and the number of plaques thus reflected the number of *E. coli* host cells in the sample. Detection was achieved down to one cell per milliliter in a 3.5-h assay. Similar coinfection strategies with the other phage mutants yielded detection limits of 10 or less target cells per milliliter in 3- to 5-h assay formats.

Phage amplification assay endpoints have been combined with several atypical detection technologies. Madonna et al. [37], for example, used matrix-assisted laser desorption/ionization mass spectrometry (MALDI-MS) to identify the molecular weight signature of the phage capsid protein. *E. coli* in pure culture was concentrated by immunomagnetic separation and then infected with phage MS2. Analysis of 1 μL of this sample by MALDI-MS was sufficient to detect the MS2 capsid protein, providing a detection limit of approximately 10^4 *E. coli* cells per milliliter in an assay time of 2 h. Guan et al. [38] combined phage amplification with a competitive enzyme-linked immunosorbent assay (cELISA) to detect *S. typhimurium* using phage BP1 and a biotinylated version of BP1. *Salmonella* cultures were incubated with wild-type BP1 phage and resulting phage supernatants added to ELISA microtiter plates coated with *S. typhimurium* smooth lipopolysaccharide (LPS) to which the phage attached. The biotinylated version of the phage was additionally added, which could be detected by the colorimetric substrate 3,3′,5,5′-tetramethylbenzidine peroxidase (TMB). If excess wild-type BP1 phage were present, due to the availability of suitable *Salmonella* host cells, then few biotinylated phage would attach and a weak colorimetric signal would be detected. If no target *Salmonella* were present, then BP1 replication would not occur and excess biotinylated phage would bind to the smooth LPS to yield an intense yellow color.

2.4 Phage Display

Phage display represents a powerful technique for creating libraries of phage each of which displays on its surface a different recognition peptide for a different biological target, thus functioning much like an antibody in its interaction with antigen [39]. Thread shaped filamentous phage such as M13, f1, and fd are engineered with user-specified nucleic acid spliced within their coat protein genes. Subsequent expression of phage genes and assembly of phage components yields mature phage particles that display foreign peptides on their surface. These so-called landscape phage libraries can embody a near indefinite number of antigen recognition sites able to serve as probes for pathogenic agents when interfaced with appropriate sensor devices. For example, Lakshmanan et al. [40] immobilized a phage library clone selective for *S. typhimurium* on to a magnetoelastic sensor surface. Binding of free *Salmonella* to the immobilized phage resulted in a change in mass that could be measured by a consequent shift in resonance frequency. In a flow-through format using artificially inoculated water or milk, detection limits of approximately 10^3 cfu mL^{-1} could be obtained within 20 min. A similar magnetoelastic sensor for the detection of *Bacillus anthracis* endospores in water yielded identical detection limits [41]. Since landscape phage function more like antibodies than phage, they do not possess host ranges based on recognition, infectability, and propagation normally associated with phage. Thus, detecting resting endospores, although not possible with a conventional phage, can be accomplished with landscape phage, as can detection of various toxins, chemicals and viral agents.

2.5 Phage-Conjugated Quantum Dots

Edgar et al. [42] developed a unique phage-conjugated quantum dot method for the detection of *E. coli*. Quantum dots are fluorescent probes consisting of colloidal semiconductor nanocrystals that exhibit high quantum yield and excellent photostability. Phage T7 was engineered to display a biotinylation peptide on its major capsid protein. Upon infection, progeny phage synthesized within the *E. coli* cell became biotinylated. After lytic release from the cell, these biotinylated phage could be detected by a streptavidin functionalized quantum dot designed to attach to and label only biotinylated phage. Thus, if no suitable host were present, then no biotinylated phage were produced and the functionalized quantum dot, having nothing to attach to, would simply be washed away. Fluorescence microscopy permitted visualization of a single quantum dot conjugated phage. Typical detection limits in laboratory mixed culture approached 10 cells mL^{-1} within an assay time of 1 h. Testing of river water samples demonstrated detection of 20 *E. coli* cells per milliliter within 1 h.

2.6 Phage-Mediated Electrochemical-Based Methods

Electrochemical phage-based assays measure changes in the potential, current, or conductivity of a solution occurring due to phage activity, such as phage infection or the release of cellular constituents after cell lysis. The SEPTIC (sensing of phage-triggered ion cascade) assay, for example, uses electrochemistry to measure microscopic voltage fluctuations occurring after a phage injects its nucleic acid into its host cell. Immediately after injection, the bacterial host will emit approximately 10^8 ions into the surrounding medium, and this release can be measured electrochemically using two thin metal film microelectrodes. Dobozi-King et al. [43] demonstrated the assay using *E. coli* as the target cell in a 5-μL nanowell sensor chip [44]. Although in these experiments *E. coli* was detected at 10^7 cfu mL^{-1}, a theoretical detection limit of 1 cfu mL^{-1} was hypothesized based on potential improvements in fluid conductivity and/or reductions in thermal noise. The assay does significantly benefit from the use of wild-type phage, so no costly or time consuming genetic manipulations are required, and, since infection is one of the earliest phage/host signals available, detection can occur within seconds.

Neufeld et al. [45] used electrochemistry to measure amperometric changes in solution due to phage-mediated cell lysis. Infection of *E. coli* by a lytic version of phage lambda ultimately lead to the release of cellular components, such as the enzyme β-D-galactosidase. β-D-Galactosidase can be measured amperometrically with a potentiostat via the addition of the substrate *p*-aminophenyl-β-D-galactopyranoside (β-PAPG) to yield the product *p*-aminophenol which is oxidized at the carbon anode. *E. coli* could be detected within 6-8 h at a detection limit of 1 cfu per 100 mL. Yemini et al. [46] used the same principle to detect *Bacillus cereus*, where lysis by phage B1-7064 instigated cellular release of the enzyme α-glucosidase, as well as *M. smegmatis* using phage D29 and the cellular release of β-glucosidase. Theoretically and advantageously, any phage/host combination can be detected using this method as long as the phage is lytic and an appropriate electrochemically detectable enzymatic marker is released by the target cell. However, a single phage would likely be insufficient to infect across the spectrum of target bacterial cells desired and false negative signals arising from cross-infections or naturally lysing cells would have to be accounted for. Neufeld et al. [47] addressed some of these concerns in their assays using a phage-encoded alkaline phosphatase enzyme that had to be delivered to the host cell in order to be expressed. Thus, only after an active infection event would the enzyme be synthesized and then later released by the cell during lysis. This assay could detect a single *E. coli* cfu mL^{-1} in less than 3 h in both pure and mixed cultures.

Chang et al. [48], based on the knowledge that growth in a microbial culture could be monitored electrochemically by measuring changes in electrical parameters occurring as complex growth media substrates were broken down into smaller highly charged molecules such as acids, hypothesized that the presence of phage within a bacterial culture, provided that suitable host cells were present, would impede culture growth and therefore directly affect growth media composition. Thus, by comparing conductance measurements between phage-supplemented and phage-free samples or between phage-specific and nonphage-specific bacterial cultures, one could easily screen samples for the presence of phage-specific pathogens. Phage AR1 and its *E. coli* O157:H7 host were used to demonstrate the technique. Pure cultures of *E. coli* O157:H7 or non-O157:H7 cells at 10^6 cfu mL^{-1} with or without phage AR1 addition were placed in test tubes fitted with platinum electrodes and conductance measurements were taken every 6 min. Resulting conductance curves could discriminate between *E. coli* O157:H7 and non-O157:H7 cultures within a 24-h period.

2.7 Surface Plasmon Resonance-Based Sensing

Surface plasmon resonance (SPR) detects changes in refractive index due to receptor/ligand binding interactions typically associated with an antibody receptor designed to associate with an antigenic bacterial pathogen ligand. Recognizing the similar capacity of phage to capture specific bacterial targets, Balasubramanian et al. [49] physically adsorbed the *Staphylococcus aureus* specific phage 12600 onto the gold surface of a commercially available handheld SPR sensor referred to as Spreeta (Texas Instruments). *S. aureus* cells were pumped through the flow-through Spreeta where they contacted and attached to the immobilized phage. Nonphage-specific bacterial cells, in this case *S. typhimurium*, simply passed through the device and were deposited as waste. The resulting change in refractive index due to phage specific *S. aureus* binding was measured, resulting in near real-time detection limits of 10^4 cells mL^{-1}. Still in developmental stages, further refinement of the assay system should drive detection limits lower.

2.8 The Phage Adenylate Kinase Assay

Adenylate kinase is an enzyme that directs the conversion of ADP to ATP and AMP (2ADP⇄ATP + AMP). When a cell bursts open after a phage infection event, adenylate kinase will be released as part of the cellular milieu. Blasco et al. [50] used this fact to create a phage-based assay whose endpoint relied on adenylate kinase driven conversion of ADP to ATP. The *E. coli* specific phage NCIMB 10359 or the *Salmonella* specific phage Newport were added to bacterial cultures, and, if suitable host cells were present, infection, lysis, and release of adenylate kinase occurred. ADP was then added to drive the adenylate kinase-mediated reaction towards the generation of ATP, and resulting ATP pools were detected using a commercially available firefly luciferase assay. Detection limits approached 10^4 cfu mL^{-1} of *E. coli* or *Salmonella* within an assay time of less than 2 h. Wu et al. [51] later optimized assay incubation times and phage concentrations to increase detection limits to 10^3 cfu mL^{-1}.

3 Concluding Remarks

With their innate ability to distinctly recognize cellular hosts, bacteriophage have realized diverse potentials in the sensing and monitoring of bacterial pathogens. Each aspect of the phage life cycle, from infection to propagation to amplified release of progeny phage, has been harnessed and manipulated to create sensitive, specific, and rapid assays for select pathogens. In addition, genetically engineered landscape phage libraries have evolved phage detection capabilities beyond living pathogens to similarly critical toxin, spore, and viral threat agents. Considering that only a handful of the massively abundant phage population has been exploited as sensors, it is clear that consolidated research efforts much more extensive than that described here will inevitably yield phage-mediated sensor interfaces covering a wide, multiplexed range of biological and biologically-derived targets. The inherent power of phage/host recognition guarantees a role for bacteriophage in the future of biosensing technologies.

References

1. Funatsu T, Taniyama T, Tajima T et al. (2002) Rapid and sensitive detection method of a bacterium by using a GFP reporter phage. Microbiol Immunol 46:365–369
2. Tanji Y, Furukawa C, Na SH et al. (2004) *Escherichia coli* detection by GFP-labeled lysozyme-inactivated T4 bacteriophage. J Biotechnol 114:11–20
3. Miyanaga K, Hijikata TF, Furukawa C et al. (2006) Detection of *Escherichia coli* in the sewage influent by fluorescent labeled T4 phage. Biochem Eng J 29:119–124

4. Namura M, Hijikata T, Miyanaga K et al. (2008) Detection of *Escherichia coli* with fluorescent labeled phages that have a broad host range to *E.coli* in sewage water. Biotechnol Prog 24:481–486
5. Oda M, Morita M, Unno H et al. (2004) Rapid detection of *Escherichia coli* O157:H7 by using green fluorescent protein-labeled PP01 bacteriophage. Appl Environ Microbiol 70:527–534
6. Awais R, Fukudomi H, Miyanaga K et al. (2006) A recombinant bacteriophage-based assay for the discriminative detection of culturable and viable but nonculturable *Escherichia coli* O157:H7. Biotechnol Prog 22:853–859
7. Meighen EA (1994) Genetics of bacterial bioluminescence. Annu Rev Genet 28:117–139
8. Ulitzur S, Kuhn J (1987) Introduction of *lux* genes into bacteria, a new approach for specific determination of bacteria and their antibiotic susceptibility. In: Scholmerich J, Andreesen R, Kapp A, Ernst M, Woods WG (eds) Bioluminescence and chemiluminescence: new perspectives. Wiley, New York
9. Kodikara CP, Crew HH, Stewart GSAB (1991) Near on-line detection of enteric bacteria using lux recombinant bacteriophage. *FEMS Microbiol Lett* 83:261–266
10. Waddell TE, Poppe C (2000) Construction of mini-Tn10*luxABcam/Ptac*-ATS and its use for developing a bacteriophage that transduces bioluminescence to *Escherichia coli* O157:H7. FEMS Microbiol Lett 182:285–289
11. Chen J, Griffiths MW (1996) *Salmonella* detection in eggs using *lux*+ bacteriophages. *J Food Prot* 59:908–914
12. Thouand G, Vachon P, Liu S et al. (2008) Optimization and validation of a simple method using P22::*luxAB* bacteriophage for rapid detection of *Salmonella enterica* serotypes A, B, and D in poultry samples. J Food Prot 71:380–385
13. Loessner MJ, Rudolf M, Scherer S (1997) Evaluation of luciferase reporter bacteriophage A511::*luxAB* for detection of *Listeria monocytogenes* in contaminated foods. Appl Environ Microbiol 63:2961–2965
14. Ripp S, Jegier P, Birmele M et al. (2006) Linking bacteriophage infection to quorum sensing signalling and bioluminescent bioreporter monitoring for direct detection of bacterial agents. J Appl Microbiol 100:488–499
15. Harms H, Wells MC, van der Meer JR (2006) Whole-cell living biosensors - are they ready for environmental application? Appl Microbiol Biotechnol 70:273–280
16. Ripp S, Jegier P, Johnson CM et al. (2008) Bacteriophage-amplified bioluminescent sensing of *Escherichia coli* O157:H7. Anal Bioanal Chem 391:507–514
17. Carriere C, Riska PF, Zimhony O et al. (1997) Conditionally replicating luciferase reporter phages: improved sensitivity for rapid detection and assessment of drug susceptibility of *Mycobacterium tuberculosis*. J Clin Microbiol 35:3232–3239
18. Sarkis GJ, Jacobs WR, Hatfull GF (1995) L5 luciferase reporter mycobacteriophages: a sensitive tool for the detection and assay of live mycobacteria. Mol Microbiol 15:1055–1067
19. Bardarov S, Dou H, Eisenach K et al. (2003) Detection and drug-susceptibility testing of *M. tuberculosis* from sputum samples using luciferase reporter phage: comparison with the mycobacteria growth indicator tube (MGIT) system. Diagn Microbiol Infect Dis 45:53–61
20. Riska PF, Su Y, Bardarov S et al. (1999) Rapid film-based determination of antibiotic susceptibilities of *Mycobacterium tuberculosis* strains by using a luciferase reporter phage and the Bronx box. J Clin Microbiol 37:1144–1149
21. Hazbon MH, Guarin N, Ferro BE et al. (2003) Photographic and luminometric detection of luciferase reporter phages for drug susceptibility testing of clinical *Mycobacterium tuberculosis* isolates. J Clin Microbiol 41:4865–4869
22. Goodridge L, Griffiths MW (2002) Reporter bacteriophage assays as a means to detect foodborne pathogenic bacteria. Food Res Int 35:863–870
23. Wolber PK, Green RL (1990) Detection of bacteria by transduction of ice nucleation genes. Trends Biotechnol 8:276–279

24. Goodridge L, Chen J, Griffiths MW (1999) Development and characterization of a fluorescent-bacteriophage assay for detection of *Escherichia coli* O157:H7. *Appl Environ Microbiol* 65:1397–1404
25. Goodridge L, Chen J, Griffiths MW (1999) The use of fluorescent bacteriophage assay for detection of *Escherichia coli* O157:H7 in inoculated ground beef and raw milk. *Int J Food Microbiol* 47:43–50
26. Kenzaka T, Utrarachkij F, Suthienkul O et al. (2006) Rapid monitoring of *Escherichia coli* in southeast Asian urban canals by fluorescent-bacteriophage assay. *J Health Sci* 52:666–671
27. Mosier-Boss PA, Lieberman SH, Andrews JM et al. (2003) Use of fluorescently labeled phage in the detection and identification of bacterial species. *Appl Spectrosc* 57:1138–1144
28. Crane DD, Martin LD, Hirsh DC (1984) Detection of *Salmonella* in feces by using Felix-01 bacteriophage and high-performance liquid-chromatography. *J Microbiol Methods* 2:251–256
29. Hirsh DC, Martin LD (1983) Detection of *Salmonella* spp in milk by using Felix-01 bacteriophage high-pressure liquid-chromatography. *Appl Environ Microbiol* 46:1243–1245
30. Stewart GSAB, Jassim SAA, Denyer SP et al. (1998) The specific and sensitive detection of bacterial pathogens with 4 h using bacteriophage amplification. *J Appl Microbiol* 84:777–783
31. Mole RJ, Maskell WOC (2001) Phage as a diagnostic – the use of phage in TB diagnosis. *J Chem Technol Biotechnol* 76:683–688
32. Pai M, Kalantri S, Pascopella L et al. (2005) Bacteriophage-based assays for the rapid detection of rifampicin resistance in *Mycobacterium tuberculosis*: a meta-analysis. *J Infect* 51:175–187
33. Favrin SJ, Jassim SA, Griffiths MW (2001) Development and optimization of a novel immunomagnetic separation-bacteriophage assay for detection of *Salmonella enterica* serovar enteritidis in broth. *Appl Environ Microbiol* 67:217–224
34. Favrin SJ, Jassim SA, Griffiths MW (2003) Application of a novel immunomagnetic separation-bacteriophage assay for the detection of *Salmonella enteritidis* and *Escherichia coli* O157: H7 in food. *Int J Food Microbiol* 85:63–71
35. Jassim SAA, Griffiths MW (2007) Evaluation of a rapid microbial detection method via phage lytic amplification assay coupled with live/dead fluorochromic stains. *Lett Appl Microbiol* 44:673–678
36. Ulitzur N, Ulitzur S (2006) New rapid and simple methods for detection of bacteria and determination of their antibiotic susceptibility by using phage mutants. *Appl Environ Microbiol* 72:7455–7459
37. Madonna AJ, Van Cuyk S, Voorhees KJ (2003) Detection of *Escherichia coli* using immunomagnetic separation and bacteriophage amplification coupled with matrix-assisted laser desorption/ionization time-of-flight mass spectrometry. *Rap Commun Mass Spectrom* 17:257–263
38. Guan JW, Chan M, Allain B et al. (2006) Detection of multiple antibiotic-resistant *Salmonella enterica* serovar Typhimurium DT104 by phage replication-competitive enzyme-linked immunosorbent assay. *J Food Prot* 69:739–742
39. Petrenko VA, Vodyanoy VJ (2003) Phage display for detection of biological threat agents. *J Microbiol Methods* 53:253–262
40. Lakshmanan RS, Guntupalli R, Hu J et al. (2007) Detection of *Salmonella typhimurium* in fat free milk using a phage immobilized magnetoelastic sensor. *Sens Actuator B Chem* 126:544–550
41. Wan JH, Johnson ML, Guntupalli R et al. (2007) Detection of *Bacillus anthracis* spores in liquid using phage-based magnetoelastic micro-resonators. *Sens Actuator B Chem* 127:559–566
42. Edgar R, McKinstry M, Hwang J et al. (2006) High-sensitivity bacterial detection using biotin-tagged phage and quantum-dot nanocomplexes. *Proc Natl Acad Sci U S A* 103:4841–4845
43. Dobozi-King M, Seo S, Kim JU et al. (2005) Rapid detection and identification of bacteria: sensing of phage-triggered ion cascade (SEPTIC). *J Biol Phys Chem* 5:3–7

44. Seo S, Dobozi-King M, Young RF et al. (2008) Patterning a nanowell sensor biochip for specific and rapid detection of bacteria. *Microelectron Eng* 85:1484–1489
45. Neufeld T, Schwartz-Mittelmann A, Biran D et al. (2003) Combined phage typing and amperometric detection of released enzymatic activity for the specific identification and quantification of bacteria. *Anal Chem* 75:580–585
46. Yemini M, Levi Y, Yagil E et al. (2007) Specific electrochemical phage sensing for *Bacillus cereus* and *Mycobacterium smegmatis*. *Bioelectrochemistry* 70:180–184
47. Neufeld T, Mittelman AS, Buchner V et al. (2005) Electrochemical phagemid assay for the specific detection of bacteria using *Escherichia coli* TG-1 and the M13KO7 phagemid in a model system. *Anal Chem* 77:652–657
48. Chang TC, Ding HC, Chen SW (2002) A conductance method for the identification of *Escherichia coli* O157:H7 using bacteriophage AR1. *J Food Prot* 65:12–17
49. Balasubramanian S, Sorokulova IB, Vodyanoy VJ et al. (2007) Lytic phage as a specific and selective probe for detection of *Staphylococcus aureus* – a surface plasmon resonance spectroscopic study. *Biosens Bioelectron* 22:948–955
50. Blasco R, Murphy MJ, Sanders MF et al. (1998) Specific assays for bacteria using phage mediated release of adenylate kinase. *J Appl Microbiol* 84:661–666
51. Wu Y, Brovko L, Griffiths MW (2001) Influence of phage population on the phage-mediated bioluminescent adenylate kinase (AK) assay for detection of bacteria. *Lett Appl Microbiol* 33:311–315

Adv Biochem Engin/Biotechnol (2010) 118: 85–112
DOI: 10.1007/10_2009_8

Published online: 19 June 2009

Cell-Based Genotoxicity Testing

Genetically Modified and Genetically Engineered Bacteria in Environmental Genotoxicology

Georg Reifferscheid and Sebastian Buchinger

Abstract Genotoxicity test systems that are based on bacteria display an important role in the detection and assessment of DNA damaging chemicals. They belong to the basic line of test systems due to their easy realization, rapidness, broad applicability, high sensitivity and good reproducibility. Since the development of the *Salmonella* microsomal mutagenicity assay by Ames and coworkers in the early 1970s, significant development in bacterial genotoxicity assays was achieved and is still a subject matter of research. The basic principle of the mutagenicity assay is a reversion of a growth inhibited bacterial strain, e.g., due to auxotrophy, back to a fast growing phenotype (regain of prototrophy). Deeper knowledge of the mutation events allows a mechanistic understanding of the induced DNA-damage by the utilization of base specific tester strains. Collections of such specific tester strains were extended by genetic engineering. Beside the reversion assays, test systems utilizing the bacterial SOS-response were invented. These methods are based on the fusion of various SOS-responsive promoters with a broad variety of reporter genes facilitating numerous methods of signal detection.

A very important aspect of genotoxicity testing is the bioactivation of xenobiotics to DNA-damaging compounds. Most widely used is the extracellular metabolic activation by making use of rodent liver homogenates. Again, genetic engineering allows the construction of highly sophisticated bacterial tester strains with significantly enhanced sensitivity due to overexpression of enzymes that are involved in the metabolism of xenobiotics. This provides mechanistic insights into the toxification and detoxification pathways of xenobiotics and helps explaining the chemical nature of hazardous substances in unknown mixtures.

G. Reifferscheid (✉) and S. Buchinger
Division of Qualitative Hydrology, Federal Institute of Hydrology (BfG),
Koblenz, 56068, Germany
e-mail: reifferscheid@bafg.de

In summary, beginning with "natural" tester strains the rational design of bacteria led to highly specific and sensitive tools for a rapid, reliable and cost effective genotoxicity testing that is of outstanding importance in the risk assessment of compounds (REACH) and in ecotoxicology.

Keywords Genotoxicity • Mutagenicity • Bacterial tester strains • SOS-response • reportergene assay • Xenobiotic metabolism • Metabolic activation

Contents

1 Introduction

Damage to the DNA is one of the most serious challenges organisms must face. Multiple more or less differentiated DNA-repair mechanisms exist in the cells of all creatures; nevertheless the risk remains that certain primary defects are transformed to mutations by misrepair, inhibition or total lack of repair that is possibly leading to the initiation of carcinogenesis. Point mutations are the cause of many human genetic diseases and there is substantial evidence that point mutations in proto-oncogenes and tumor suppressor genes of somatic cells are involved in tumor formation in humans and animal experiment [1]. In environmental toxicology the exceptional position of genotoxicity assessment compared to general toxicity is – in simple terms – based on two postulates: (1) the absence of a threshold value for genotoxic carcinogens, i.e. the absence of a no effect concentration and (2) the irreversibility of mutations. The toxicological endpoint "DNA-damage" is embedded in the triangle "genotoxicity–mutagenicity–carcinogenicity." From a statistical viewpoint, the clear relationship between primary DNA-damage, mutagenicity and carcinogenicity becomes obvious. Furthermore, multiple secondary DNA-damages affect the state of health of an organism beyond carcinogenesis, e.g. promotion of aging processes or reduction of lifespan.

Only a limited number of genotoxins that are found in the environment can be identified by their chemical nature. Serious problems may arise when toxicologically not sufficiently evaluated chemicals are detected in drinking water resources or during drinking water processing. In such cases a battery of quickly responding test systems combined with the possibility to align chemical analyses with genotoxicity databases is necessary.

Concepts for the systematic evaluation of the genotoxicity, mutagenicity and carcinogenicity of pharmaceuticals have already been established decades ago. In this context bacterial test systems belong to the basic line of test systems due to their easy realization, rapidness, broad applicability, high sensitivity and good reproducibility. Additionally, the specificity of the various test strains provides useful information on the types of mutations that are induced by genotoxic agents.

Two basic principles for the detection of DNA-defects in bacteria have been established: the reverse mutation principle for detecting point mutations and second, the measurement of primary DNA-damage by induction of the bacterial SOS-response. However, guidelines for the assessment of pharmaceuticals consistently recommend the reverse mutation tests to be part of a standard test battery as this test principle has been shown to detect relevant genetic changes and the majority of genotoxic rodent carcinogens [2].

Initially, bacterial test systems were used for the assessment of single chemicals. In recent years numerous applications on environmental matrices have been published. Perceiving the presence and potency of genotoxins in the environment originating naturally or from diffuse or punctual sources like industrial and municipal effluents, an OSPAR expert group addressed the necessity of genotoxicity testing for the evaluation of waste water within whole effluent assessment [3]. For the measurement of environmental samples, two ISO standards on bacterial genotoxicity assessment have been developed so far [4, 5].

Bacterial indicator tests as well as reverse mutation tests utilize prokaryotic cells, which differ from eukaryotic cells with respect to uptake and metabolism of chemicals, chromosome structure and DNA repair processes. Tests conducted *in vitro*, generally require the use of an exogenous source of metabolic activation (S9-mix, see below). However, in vitro metabolic activation systems cannot sufficiently mimic the mammalian in vivo conditions. Bacterial tests therefore do not provide direct information on the mutagenic and carcinogenic potency of a substance in mammals [1]. On the other hand, no single test system – prokaryotic or eukaryotic – is capable of detecting all relevant genotoxic agents. The usual approach to carrying out a comprehensive assessment of genotoxins is to apply a battery of complementary in vitro and in vivo systems. The general features of a recommended standard test battery (for testing of pharmaceuticals) are described in detail in the ICH S2B genotoxicity guidance paper [6]. However, even a standard test battery cannot cover all "special cases" so that scientifically justified deviations from standard procedures are occasionally necessary.

This chapter summarizes current knowledge on bacterial genotoxicity and mutagenicity tester strains and systems with special focus on recent developments. The authors are aware that a complete discussion of all aspects of genotoxicity is beyond the scope of this article.

2 Bacterial Test Systems Based on the Reverse Mutation Principle

For decades bacterial test systems have represented the first line for genotoxicity and mutagenicity assessment of chemicals and in recent years also for environmental purposes.

The era of bacterial mutagenicity testing started in the late 1960s and early 1970s with Bruce Ames and Phil Hartman, who worked on the histidine regulation in *Salmonella typhimurium* using histidine auxotrophic mutants. Phil Hartman screened over 1,000 clones for mutants with low background level for spontaneous reversion frequencies and optimal responses to particular known mutagens [7]. In 1972 Bruce Ames published one of the first papers on mutagenesis in bacteria using precursor strains of the standard tester strains that are applied down to the present day [8]. The basis of mutagenicity testing was a set of bacterial strains defective in histidine synthesis but highly sensitive to chemical- or radiation–induced mutagenesis resulting in the reversion to prototrophy, i.e., independence from histidine supplementation. In this system the target gene sequence for mutagenesis is the histidine operon consisting of several consecutive, coregulated genes (*hisLGDCBHAFI*). The long and complex pathway of histidine synthesis is connected with nucleic acid biosynthesis.

Histidine auxotrophic *Salmonella* bacteria are unable to proliferate in a growth medium lacking histidine, but they can specifically revert to the original genetic wild type. Alternatively, they can recover prototrophy by reconstituting a base sequence that allows the production of functioning enzymes that are necessary for histidine synthesis. Sensitive sites for reversion (hot spots) were selected from three genes of the his operon: *hisC*, coding for the histidinol-phosphate aminotransferase, *hisD*, a bifunctional histidinal and histidinol dehydrogenase, and *hisG*, coding for a histidinol-phosphate aminotransferase.

The ingenious original test design – application of the test bacteria on minimal agar plates supplemented with a trace of histidine in order to allow some cell divisions for transforming primary DNA-damage into mutations – has been maintained with some modifications up to date.

A first milestone for the establishment of bacterial mutation test systems was Ames' paper entitled "Carcinogens are mutagens: a simple test system combining liver homogenates for activation and bacteria for detection" [8]. In this paper he described a simple, inexpensive, and extremely sensitive test for the detection of carcinogens as mutagens. From the very beginning the test combined the use of a rat or human liver homogenate for carcinogen activation (supplying mammalian metabolism) and a set of *Salmonella* histidine auxotrophic mutants for the detection of base-pair substitutions (strain TA1535) or various kinds of frameshift mutations (TA1536, TA1537, and TA1538) [9]. Later on the *Salmonella* microsomal assay was mostly called "Ames test" for short after the pioneer of classical bacterial mutagenicity testing.

The first years of bacterial mutagenicity testing showed a high correlation between carcinogenicity and mutagenicity. In a study by McCann and Ames [10], 90% of the carcinogens were mutagenic in the Ames test, including almost all of the known human carcinogens that were tested. This statement had to be relativized more and more in later years due to (1) the highly artificial test design by supplementation of S9-mix leading to a worst-case scenario that in fact detects a wide variety of bacterial genotoxins and mutagens but shows false-positive results with respect to carcinogenicity and (2) the intrinsic inability to detect nongenotoxic carcinogens.

The basis for sensitive mutagenicity testing were genetically manipulated bacterial cells possessing a special genotype in order to increase their sensitivity to chemical mutagens.

The mutation *rfa* causes partial loss of the lipopolysaccharide barrier that coats the surface of the cells, thus increasing their permeability to large and lipophilic molecules like polycyclic aromatic hydrocarbons. Another mutation (*uvr*) causes deficiency in the DNA excision repair system. Both mutations have significantly enhanced the mutagen sensitivity of the Ames test. Since the strains bearing deletions of *uvr* (*uvrB–bio–gal*) and *rfa* have been independently isolated, they are not isogenic, although there is no evidence that these differences are of any significance in the Ames test [11]. However, there is evidence that the genetic background of base substitution strain TA 100 appears to be more sensitive to cytotoxic effects of chemicals compared to the frameshift strain TA 98 [11].

A second quantum jump was the introduction of the so-called R-factor (resistance transfer) plasmid pKM101 carrying an antibiotic resistance gene (ampicillin) and the *mucAB* genes that allow error-prone DNA-synthesis and thereby facilitating the induction of mutations [12, 13]. The plasmid was introduced into the strains TA 1535 now called TA 100 and into TA 1538 now called TA 98. Strain TA 1535 and TA 100 are specific for G to A transitions (reversion to the wild type), but as well detect G to T and G to C transversions. According to [14] 60–80% of all substitution events are G to A transitions and G to T transversions. Strain TA 98 detects both, +1 and −2 frameshift mutations. A section of at least 60 base pairs ahead the insertion site can act as the target sequence for frameshift mutations [15].

McCann et al. [12] presented evidence that mutagens that become more effective in the presence of the R factor plasmid pKM101 act via an error-prone recombinational repair. Shortly thereafter, it became obvious that all these effects depend on the *recA+ lexA+* genotype. The *recA/lexA*-dependence of the plasmid-mediated DNA-repair and mutagenesis suggested an interaction with the cell's inducible error-prone repair system [16], that is regulated by the bacterial SOS-response. It could also be shown that mutagenesis by UV and many chemicals in *Escherichia coli* require the products of the *umuDC* operon or the analogous plasmid-derived operon *mucAB* [17]. The products of *mucAB* and the *E. coli umuDC* operon share 52% homology at the nucleotide level [18].

With the construction of the tester strains TA 97 (similar to TA 1537 plus plasmid pKM101; [19]) for the detection of −1 frameshift events in a C-run and the excision repair proficient strain TA 102 for the detection of transitions and transversions especially in connection with oxidative mutagens and DNA–protein and DNA–DNA crosslinking agents [19], the basic set of Ames tester strains was complete. Strain TA 102 is the only genetically engineered strain in this set. It differs from previous tester strains due to the introduction of the mutation via a multicopy plasmid (termed pAQ1), so that approximately 30 copies of the mutant gene are available for back mutation. Consequently, the number of spontaneous reversions is comparatively high. Further on the strain contains A:T base pairs at the site of the mutation in contrast to the other tester strains that detect mutagens damaging G:C base pairs. The strain is especially applicable to bifunctional genotoxins (crosslinkers) and gyrase inhibitors that are highly cytotoxic and only express their mutagenic potential in DNA repair proficient cells.

In case of the tester strain TA 102, substances that increase the number of pAQ1 plasmids per cell (e.g., antibiotics like tetracycline and chloramphenicol) apparently also increase the amount of mutant colonies. Actually they do not increase the mutation frequency per gene, but only enhance the number of the genes at which mutations can occur [20]. In general, the consideration of the molecular processes of mutagenesis is important in order to avoid the misinterpretation of test results.

According to the OECD guidelines for the testing of chemical substances, usage of at least five test strains (TA 1535, TA 1537 or TA 97, TA 98, TA 100 and TA 102) is recommended for a comprehensive detection of various chemical mutagens. As an alternative to TA 102, the repair proficient *E. coli* WP2 strain may be used (Table 1).

The combination of TA 98 and TA 100 appears to be an optimal reduced strain set for time and labor efficient testing, that still covers a broad range of possible mutagens. The ISO standard for water testing (ISO 16240) [5] uses this reduced strain combination. The conventional Ames test described in OECD guidelines and the ISO standard is performed as a plate incorporation assay on agar plates. Versions with complete and reduced sets of strains have been used in plenty of cases to evaluate the mutagenicity of single substances and environmental samples, in the latter case mostly in combination with specific extraction procedures. The PubMed

Table 1 Basic tester strains for the detection of mutations in bacteria[a]

Strain	Description	Specificity	Source or reference
S. typhimurium	his deficiency		
TA1535[b]	*hisG46*	bps	[9]
TA1537[b]	*hisC3076*	−1 frs in a C-run,	[9]
TA1538	*hisD3052*		[9]
TA97[b]	*hisD6610* (pKM101)	−1 frs in a C-run, error prone repair	[21]
TA98[b,c]	*hisD3052* (pKM101)	−2 frs, error prone repair	[12]
TA100[b,c]	*hisG46* (pKM101)	bps, error prone repair	[12]
TA102[b,d]	*hisG428* (pKM101, pAQ1)	bps, error prone repair, excision repair proficient	[21]
E. coli	trp deficiency		
WP2[b]	AT as critical mutation in	bps	[22]
WP2 (pKM101)[b]	the *trpE* gene	bps, error prone repair; excision repair proficient	[23]

[a]The bacterial strains listed in this and all following tables represent a selection of frequently used genetically modified and/or genetically engineered tester strains. It is not intended to list completely all existing types of strains in this essay

[b]Tester strain recommended in the OECD guideline for testing of chemicals - bacterial reverse mutation test

[c]Tester strain recommended in ISO16240

[d]Genetically engineered tester strain

bps Base pair substitution; *frs* Frameshift mutation; *his* Histidine; *trp* Tryptophan

database (http://www.ncbi.nlm.nih.gov/pubmed/) alone contains thousands of entries for the Ames test that is performed in the last few decades all over the world. Most mutagenicity data have been generated with the conventional plate incorporation test. Several test-versions aiming to increase test sensitivity or reduce required sample amount have been developed in recent years, e.g., microsuspension and preincubation test versions [24–27] and fluctuation test systems [28–30].

3 Base Specific Tester Strains

Multiple modes of reversion characterize each Ames tester strain. However, the strains of the conventional Ames test are not fully diagnostic for the type of base-pair substitution caused by the mutagen. This drawback could be overcome by developing a set of six different histidine auxotrophic *Salmonella* strains, each specific for a certain substitution allowing the detection of patterns that indicate which of the six possible base substitutions (Table 2) are induced by a given mutagen [31]. The strains have considerably lower spontaneous reversion frequencies compared to the conventional tester strains and can thus be applied in a mixture followed by a classification of the type of mutation with the individual strains. An international comparative study using the tester strains TA 98 and the base specific substitution strains TA7001 – TA7006 as a "equimolar" mixture instead of only the strain TA 100- this approach (called the Ames II mutagenicity assay) was recommended as an effective screening alternative to the standard Ames test-proved a high concordance between the two test systems but required less test material and labor.

Several other bacterial systems besides the histidine reversion assay that detect all possible base substitutions without further genetic or biochemical analysis have been reported. Among them is an *E. coli* system bearing different mutations at the same coding position in the *lacZ* gene (Table 2), which specifies a glutamic acid residue at position 461 that is located in the active site of the beta-galactosidase [32]. Each strain is Lac− and reverts to Lac+ only by restoring the glutamic acid codon. Each transition and transversion is simply monitored by the Lac− to Lac+ frequency after incubation on a lactose minimal medium for selection. An alternative reversion test principle using base specific tester strains coupled with reporter gene detection of reversions is described in the following section.

Table 2 Base specific tester strains for mutagenicity testing

Strain	Description	Specificity	Source or reference
S. typhimurium			
TA7001-TA7006	*hisG1775, C9138, G9074, G9133, G9130, C9070*	All possible bps in the *his* operon	[31]
E. coli			
CC101-106	*lacZ- mutation at residue 461 of the beta-galactosidase*	All possible bps in the *lacZ* gene	[32]

It is worth noting that, beside the reversion principle, several forward mutation test systems have been developed. In contrast to a reversion, a forward mutation converts a wild-type allele to a mutant allele. One test is based on the gain of 5-fluorouracil resistance by the mutation event using a strain of *S. typhimurium* that is derived from the Ames strain TA 100. Of 25 mutagens and 20 nonmutagens tested, the results correlated to 100% with those obtained by the battery of standard Ames reversion strains. This assay proved to be really sensitive but takes more than 3 days to be completed [33]. A second system – the arabinose resistance test – with *S. typhimurium* (Ara test) is a forward mutation assay which selects a single phenotypic change (from L-arabinose sensitivity to L-arabinose resistance) in a unique tester strain (an *araD* mutant) [34].

4 Combination of the Reversion Test Principle with Reporter Gene Detection

In an effort to combine the advantages of mutagenicity test systems with those of indicator tests [35] developed a new reversion-based mutagenicity test that utilizes reporter-gene-mediated detection of revertants in liquid culture. In this system, called the "mutagen assay," a reversibly knocked-out TEM-1 class A beta-lactamase gene coding for ampicillin resistance acts as the selection marker. In order to detect mutagens quickly and sensitively, the system utilizes a series of plasmids that contain the mutated ampicillinase gene and the *mucAB* operon [30]. Deactivating mutations in the ampicillinase gene include frameshifts integrated into repetitive GC-sequences and G-runs known to be mutagenic hot-spots, and all possible base-pair substitutions inserted in or around the active site of the beta-lactamase (Table 3). An advantage of the MutaGen assay over other reversion tests is

Table 3 Genetically engineered strains for the detection of mutations in bacteria based on reporter gene detection

Strain	Specificity[a]	Source or reference
S. typhimurium		
TA1535	bps	
3S	G:C →A:T	[35]
18S	T:A →A:T	[30]
19S	C:G →A:T	[30]
20S	T:A →G:C	[30]
21S	C:G →G:C	[30]
22S	A:T →G:C	[30]
	frs	
4S	−2 frs	[35]
6S	−1 frs	[35]

[a]Indicating reversion to the proficient strain
bps base pair substitution ; *frs* frameshift mutation

that nonreverted bacteria disappear due to lytic death in presence of ampicillin [36, 37], whereas beta-lactamase-proficient mutants continue to grow. Due to a nutrient-optimized complete growth medium, nonspecific mutations in genes essential for optimal growth are of minor relevance. The short generation time in complete medium allows detection of mutations within only one working day.

A crucial requirement for performance as a liquid-incubation assay is stringent control of the reporter gene to ensure a negligible signal background during preculture and preincubation. This was achieved by integration of the *tet* repressor gene downstream of the constitutively expressed lactamase gene. The repressor protein tightly suppresses expression of the *tetA* promoter-regulated *lacZ* reporter gene [38, 39]. The *tetA* promoter is inducible by anhydrotetracycline.

5 Genetically Engineered Promoter–Reporter Gene Fusions for Measurement of Genotoxicity Based on the Bacterial SOS-System

In the middle of the 1970s a hypothesis was proposed that *E. coli* possesses a pleiotropic response (the SOS response) to treatments that damage DNA or inhibit DNA replication, triggering an inducible DNA repair system (SOS repair, Table 4) which is also responsible for induced mutagenesis [40]. SOS repair is induced following damage to DNA. It requires de novo protein synthesis and several genetic functions of which the best-studied are *recA* and *lexA* of *E. coli*. The products of the *lexA* and *recA* genes regulate the cellular response to DNA damage whereby the *lexA* gene product represses its own promoter [41], the *recA* gene [42] and all other

Table 4 Genes, induced during SOS-response of E. coli

Gene	Encoded protein and/or function in DNA repair
Genes with known function	
polB (*dinA*)	Polymerization subunit of DNA-polymerase I; restart of replication after recombination repair
uvrA[a] } *uvrB*	UvrA- and UvrB-subunits of ABC Excinuclease
umuC[a] } *umuD*[a]	DNA-polymerase V
sulA[a]	Inhibition of cell division
recA[a]	RecA-protein, initiation of SOS response; responsible for error prone repair and recombination repair
lexA[a]	LexA-repressor, inhibition of transcription of all SOS-genes
dinB	DNA-polymerase IV
cda[a]	Inhibition of protein synthesis by cleaving tRNA(Arg)
Genes with unknown function	
dinF	

[a]Promoters used for the construction of genotoxicity tests based on the SOS response

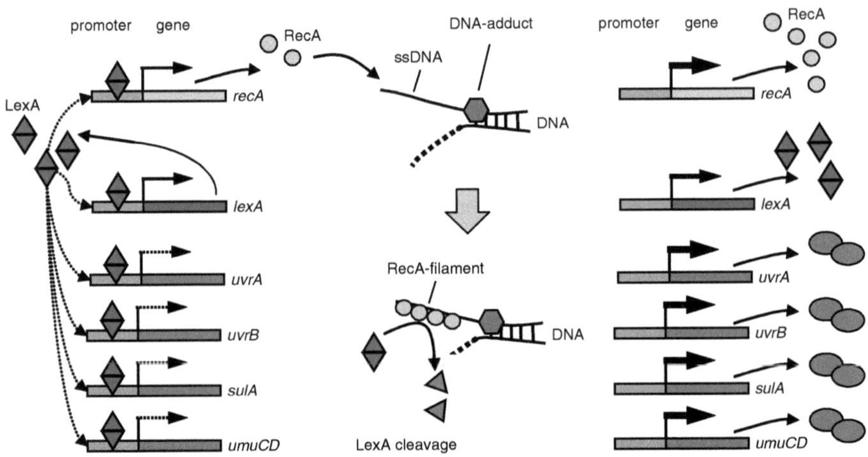

Fig. 1 Simplified molecular biological model of the SOS response in *E. coli*

SOS responsive genes [43] (Fig. 1). Single-stranded DNA–RecA complexes (RecA-filament) are build in consequence of DNA-damage and act as a coprotease in a LexA self-cleaving mechanism [42, 44, 45]. In consequence, the SOS genes are derepressed and they express various functions like inhibition of cell division (*sfiA*; [46]) and nucleotide excision repair [47] that are believed to aid cell survival in induced cells (Table 4). The *cda* gene (colicin D activity), located on the *E. coli* plasmid ColD [48] encoding for the cytotoxic protein colicin D that is secreted by certain strains of *E. coli* is a somehow exceptional case of the SOS response. Colicin D activity is known to kill target cells by cleaving tRNA(Arg) [49, 50].

Based on the SOS-regulon, Quillardet et al. [51] constructed one of the first bacterial genotoxicity assays as a test system for primary DNA-damage (indicator test). The test is based on the induction of the *sfiA*(*sulA*)–gene, whose level of expression was monitored by means of a *sfiA*::*lacZ* operon fusion in *E. coli*.

The response was rapid (a few hours), and did not require survival of the tester strain as is necessary for the Ames test. The assay simply involved the colorimetric measurement of beta-galactosidase activity under the control of the *sfiA* promoter as an indirect measure of *sfiA*-induction in consequence of primary DNA-damage.

A number of known genotoxic substances were comparatively tested with respect to their SOS inducing potency, determined by the SOS chromotest, and second to their mutagenic potency, determined by the *Salmonella* assay (mutatest). This study revealed a striking correlation between the SOS inducing potency and the mutagenic potential over a wide dynamic range that span more than seven orders of magnitude. From a theoretical standpoint the results suggested that the mutagenic potency measured in the Ames test reflects the level of SOS-induction and, furthermore, that most genotoxins are inducers of the SOS response in bacteria (Table 5).

Later on, methods for the detection of environmental genotoxins based on the activation of the bacterial SOS-system have been repeatedly developed. The expression

Table 5 Genetically engineered strains for the detection of primary DNA-damage in bacteria based on the SOS-response

Strain	Description	Source or reference
S. typhimurium		
TA1535 pSK1002[a]	*hisG46, gal, del(chl, uvrB, bio), rfa,* plasmid pSK1002 *lacZ* under the control of the *umuDC* promoter	[52]
E. coli		
PQ37		[51]
C600(pLS-1)	*luxCDABFE* under control of the *cda* promoter	[53]
MG1655(pANO1)	*gfp* under control of the *cda* promoter[a]	[54]

[a]Also *recA*, *umuDC*, and *sulA* SOS-GFP vectors were constructed in this study

properties of a SOS gene (promoter)–reporter gene fusion product depends on the specific LexA-binding properties of its promoter, which are determined by the LexA binding sequence (SOS box). The SOS boxes show a considerable homology but are distinct from gene to gene. Consequently, the response of an SOS-based genotoxicity assay is significantly influenced by the promoter used [54]. Several SOS promoters have been integrated in the SOS test systems. Among them are *sulA* (*sfiA*), *umuDC*, *recA*, *recN*, *uvrA*, *alkA*, and *cda*. According to [54] the *cda* promoter is superior with regard to both response and sensitivity, due to the combination of a very low basal promoter activity and a strong response to induction influenced by the specific LexA binding properties of the promoter [54]. The tested *sulA* and *recA* constructs showed a high specific signal after induction as well, but were less sensitive due to the high background levels of reporter gene expression (*gfp*, see below). The background level of the *umuCD'* promoter was low but promoter activity was comparatively weak after induction. These findings were correlated with the structure of the promoter with respect to (1) sequence, number and arrangement of the LexA-binding sites and (2) the −10 and −35 sequences determining the "strength" of a promoter. The *umuDC'* promoter contains two LexA binding sites but shows a strong derivation from the consensus sequences at −10 (TATAAT) and −35 (TTGACA) resulting in (1) low background activity of the promoter due to strong inhibition by LexA and (2) comparatively weak activity after induction, possibly due to weak binding of the RNA-polymerase. The *recA* promoter has nearly consensus sequences at positions −10 and −35 but only one binding site for LexA with a strong deviation compared to the consensus of the LexA-binding site (CTGTATATATATACAG). Consequently, it shows high background levels of reporter gene expression combined with strong activity after induction. Like the *recA'* promoter the *sulA'* promoter has only one LexA binding site together with a high sequence similarity to the consensus at −10 and −35. The *cda* promoter combines consensus like sequences at −10 and −35 with two LexA binding sites that are merged together and located at the start of transcription. This structure of the promoter region of *cda* might explain the good performance of its promoter in a SOS dependent reporter gene assay. All these test systems are based on the fusion of an SOS-controlled promoter with a reporter gene that can be measured colorimetrically, luminometrically

Table 6 Frequently used DNA damage inducible promoters and reporter genes[a]

SOS-promoters	Reporter genes	Principle of measurement
umuDC	lacZ (galactoasidase)	Colorimetric
sulA (sfiA)	luxCDAB*(F)E* (luciferase)	Luminometric
recA	gfp (green fluorescence protein)	Fluorimetric
recN	phoA (alkaline phosphatase)	Electrochemical
uvrA		
cda		

[a] SOS-promoters and reporter genes can be fused in any combination. Prinicple of measurement depends on reporter gene.

or by emission of fluorescence light (Table 6). Among the colorimetric reporter gene tests are the Rec-lac test [55] and the *umu*-test [52] using a *umuDC*::*lacZ*-fusion.

The Vitotox test uses luminometric detection by a *recN*::*luxCDABE* construct [56]. Several other luminescent approaches use *recA*::*lux*, *uvrA*'::*lux* or *alkA*'::*lux* fusions [57, 58]. Also, [53] developed a luminescent SOS reporter gene assay named SOS *lux* test. This bioassay is based on a recombinant plasmid which carries the *luxCDABFE* genes of *Photobacterium leiognathi* downstream of a truncated cda gene with a strong SOS promoter. The authors expected to use this kind of assay for in situ and continuous detection of genotoxins.

An attractive alternative for a reporter element is the use of the green fluorescent protein, GFP, the gene of which has been cloned from the jellyfish *Aequorea victoria* first [59]. Later on, additional genes coding for different fluorescent proteins were isolated or genetically engineered in order to adapt the characteristics of the protein according to scientific demand [60–64]. An advantage of the GFP is that it is functioning without further need of supplemented cofactors or enzyme substrates allowing the development of on-line detection methods in principle. Various *gfp*-genes were fused to different SOS-sensitive promoter elements like *umuDC*, *recA*, *cda* and others [54, 65–67]. In comparative studies like [65] the performance of fluorescent proteins as reporters were tested, e.g., against *lux*-based assay systems. In general the GFP approach has been characterized as inferior to enzyme-based reporters like *lacZ* in terms of sensitivity and response kinetics [68, 69]. Several attempts were undertaken in order to overcome these drawbacks, namely use of chimeric promoters [66], higher intensity GFP variants (mutant 3) [67] or the simultaneous induction of a group of different *gfp*-genes [68]. Another approach in order to circumvent limitations of the GFP as a reporter element is reported by Norman et al. [70] who utilized a combination of a *cda*::*gfpmut3* fusion with signal detection by flow cytometry. This invention led to the development of the *cda* GenoTox assay that showed a high sensitivity to *N*-methyl-*N*′-nitro-*N*-nitrosoguanidine [71].

All test versions described above depend on the induction of LexA controlled genes of the SOS-regulon during genotoxic stress (Fig. 2).

Compared to the *Salmonella*/microsome assay, SOS tests seem to have some practical advantages. The tests are simple, the results are obtained within a few hours and only one basic strain is required for the detection of a wide range of genotoxins.

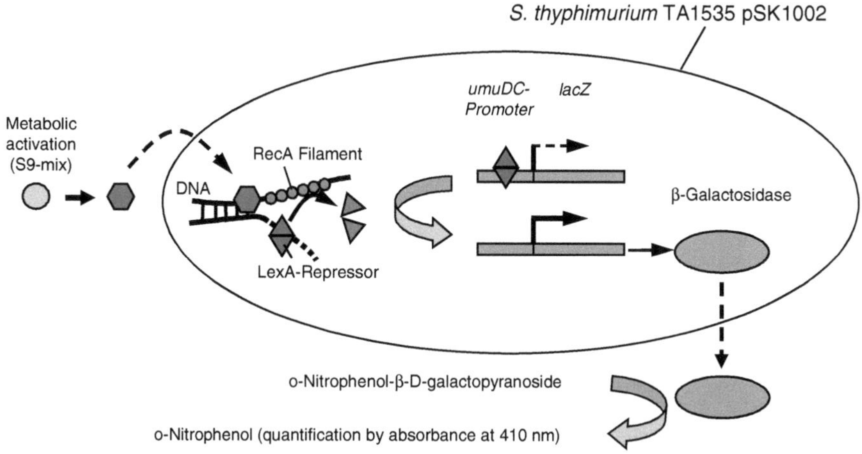

Fig. 2 Basic principle of genotoxicity-testing via SOS-induction by the fusion of a SOS-sensitive (LexA-repressed) promoter to a reporter gene (e.g., *lacZ*) illustrated by the *umu* test

However, despite their ability to detect genotoxins and potential mutagens quickly, it is worth mentioning that detection of SOS-induction cannot simply be equated with mutagenicity. Some genotoxins are mutagenic independently from the error-prone repair pathway responsible for SOS mutagenesis and therefore cannot be detected by SOS-tests. One promiment example is the carcinogenic substance cyclophosphamide that induces base pair substitutions in Ames tester strains TA 1535 and TA 100. Von der Hude et al. found that SOS-tests like the SOS-chromotest are less sensitive in detecting genotoxins that only induce frameshift mutations in the Ames frameshift tester strains [72].

6 Xenobiotic Metabolism and Metabolic Activation

Organisms are permanently exposed to an immense number of substances with varying physico-chemical properties. According to their lipophilicity and chemical structure, xenobiotics accumulate to a greater or lesser extent in lipid rich tissues and membranes affecting their normal function. In order to enhance water solubility and to allow efficient elimination of substances, during evolution a highly sophisticated and variably regulated enzyme system for the metabolism of xenobiotics developed (Fig. 3). In principle the associated biotransformation reactions can be subdivided into two sequential pathways. In the so-called phase I pathway, cytochrome P450 enzymes, also known as mixed functional oxygenases (MFO), generate oxidation products, mostly hydroxylated metabolites.

Phase I intermediates are appropriate substrates for phase II enzymes that use functional groups like hydroxyl, carboxyl, or sulfhydryl moieties created in phase I for conjugation with molecule moieties arising from the intermediate metabolism.

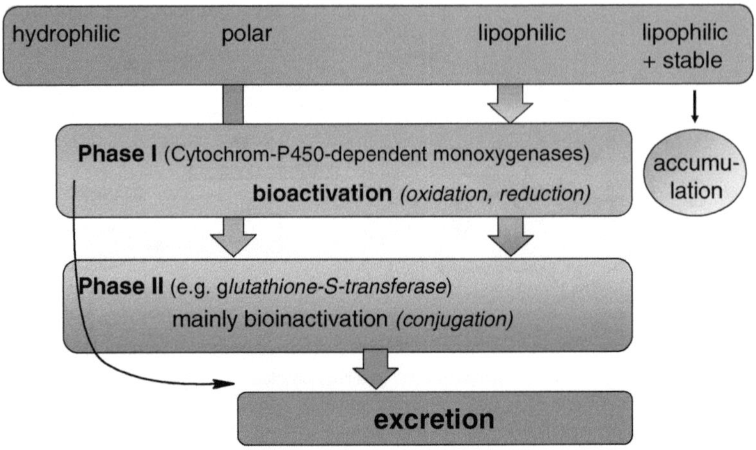

Fig. 3 Pathways of xenobiotic metabolism

Particularly glutathione, UDP-glucuronic acid, phosphoadenosyl-phosphosulfate (PAPS), acetyl coenzyme A, *S*-adenosyl-methionine, glycine and glutamine serve as conjugation partners. In sum these reaction steps facilitate the renal excretion of xenobiotics and enhance the body clearance of hydrophilic compounds. On the other hand, numerous genotoxic substances display their DNA-reactive potential only after metabolic activation by phase I or phase II reactions. A fact that underlines the immense importance of the implementation of adequate biotransformation in the context of genotoxicity testing. As an example, in phase I metabolism metabolic activation occurs by the intermediate generation of electrophilic carbenium ions after epoxide formation of polycyclic aromatic hydrocarbons like benzpyrene. These electrophilic metabolites specifically attack the extracyclic N2 of guanine leading to bulky adduct formation (Fig. 4). In addition such metabolites also covalently bind to other cellular macromolecules like RNA, proteins, and lipids.

Primarily phase II enzymes exhibit detoxification properties and inactivate compounds prior to excretion. However, phase II metabolism can also lead to products with higher toxicological activity. The glutathione dependent activation of dihaloalkanes to electrophilic episulfonium ions may be taken as an example [74–76].

Biotransformation competence and capacity is not equally present in all cells of an organism and is highly variable among creatures of different taxonomical groups. It is well known that species differ in catalytic activities of enzymes involved in the metabolism of xenobiotics. Furthermore, enzyme polymorphisms lead to interindividual differences in carcinogen metabolism. As an example, species-specific isoforms of CYP1A, −2C, −2D and −3A show appreciable interspecies differences in terms of catalytic activity [77]. In order to define doses in risk assessment which will not result in toxic effects, conventional approaches use safety factors for interspecies extrapolation and extrapolation from the "mean" subject to the general population, including sensitive subgroups. Traditionally, a default factor of 10 has been used to account for interspecies variation. To address toxicokinetic and toxicodynamic aspects it is proposed that this factor is subdivided into subfactors [78].

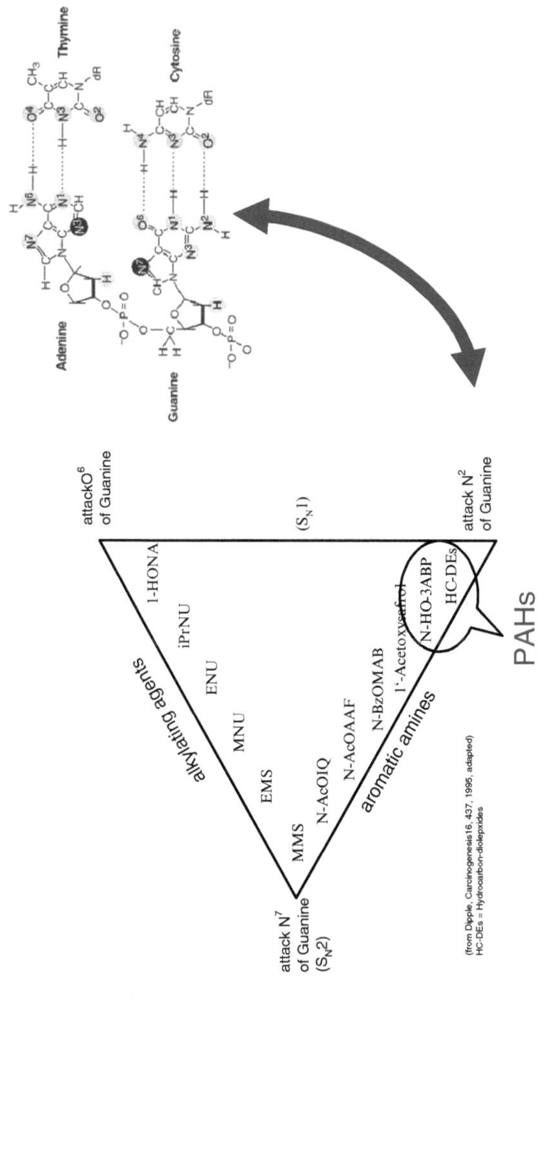

Fig. 4 Specificity of selected DNA damaging chemicals (adapted from [73])

Since the work of Malling [79] metabolizing liver preparations from induced rodents have been widely used in genotoxicity and mutagenicity studies in order to mimic the important aspect of the mammalian toxification process of potential genotoxic compounds. The liver S9-fraction became an essential component in standardized in vitro laboratory assays.

For CYP450 enzyme induction, rats (e.g., Sprague-Dawley) are simultaneously treated with phenobarbital intraperitoneally and β-naphthoflavone orally in appropriate vehicles. In case of polychlorinated biphenyl induction, the rats receive an intraperitoneal injection of a polychlorinated biphenyl (e.g., Aroclor 1254) dissolved in an appropriate vehicle. Livers are removed under sterile conditions, washed and homogenized. Finally the homogenate is centrifuged at 9,000 g in order to obtain the S9 fraction.

For external metabolic activation in the *in vitro* tests, bacteria are exposed to chemicals or environmental samples in the presence of S9 homogenate which is applied as a mixture (S9-mix) of enzymes and the cofactors glucose-6-phosphate and NADP. In general, the use of S9-mix aims to a maximum metabolic activation of chemicals by phase I enzymes. Such a worst case scenario tries to indicate any potential genotoxicity of chemicals and thus might overestimate the real *in vivo* risk.

Due to the immense variation range of chemical structures and possible biotransformation reactions the application of a standard S9-mix can only be a compromise. For this reason the conditions for optimal metabolic activation (and deactivation) of specific chemicals in vitro are not given in either case. To complicate matters further, the predominantly detoxifying phase II enzymes in the liver homogenate, though stable for a certain time, lose activity with continuing depletion of educts that are needed for the conjugation reactions (e.g., glutathione). Furthermore, external application of S9-mix can be problematic when highly polar metabolites arise that cannot cross the bacterial cell wall to enter the cells [80].

Thus for in vitro approaches the external metabolising system should be adapted to the sample to be tested considering all available information on the chemical identity of the sample.

While external metabolic activation systems were already developed in the 1960s and 1970s, and the fundamentals of metabolic activation were discovered during the 1980s, in recent years several attempts have been made to integrate genes belonging to phase I and phase II metabolic pathways into bacteria by genetic engineering. Tests with specifically engineered bacteria have been developed for both, mutagenicity assays and tests based on the SOS system.

7 Genetically Engineered Bacterial Strains with Cytochrome P450 Enzymes

The cytochrome P450 proteins (CYP) formerly called major phase I metabolic enzymes belong to the large superfamily of heme-thiolate proteins that catalyze the oxidation of various endogenous and exogenous compounds. Mammalian cytochrome

Table 7 Major classes of human cytochrome P450s

Steroids	Xenobiotics	Fatty acids	Eicosonoids	Vitamins	Unknown
1B1	1A1	2J2	4F2	2R1	2A7
7A1	1A2	4A11	4F3	24A1	2S1
7B1	2A6	4B1	4F8	26A1	2U1
8B1	2A13	4F12	5A1	26B1	2W1
11A1	2B6		8A1	26C1	3A43
11B1	2C8			27B1	4A22
11B2	2C9				4F11
17A1	2C18				4F22
19A1	2C19				4V2
21A2	2D6				4X1
27A1	2E1				4Z1
39A1	2F1				20A1
46A1	3A4				27C1
51A1	3A5				
	3A7				

P450s were originally discovered in rat liver microsomes. They are primarily membrane-associated proteins, located either in the inner membrane of mitochondria or in the endoplasmic reticulum of cells. CYP450 enzymes were identified in the mucosa of the gastrointestinal tract and in the blood brain barrier as well. MFOs are responsible for the metabolism of drugs and toxic compounds as well as metabolic products present in liver cells. Furthermore, they play important roles in estrogen and testosterone synthesis.

Most CYPs can metabolize multiple substrates, and many can catalyze multiple reactions, a fact that accounts for their central importance in metabolizing the extremely large number of different compounds. In humans alone, 57 different types of CYP-enzymes are known today (Table 7). About one quarter of them are considered to be involved in the metabolism of xenobiotic chemicals [81].

CYPs 3A4(+5), 2C19, 2C9, 2D6, 1A2, 2E1, 2B6, and 1A1 have been identified as main contributing enzymes to drug metabolism (Fig. 5). The broad substrate specificity of these MFO includes pharmaceuticals, pesticides and pollutants including promutagens and procarcinogens.

The most common reaction catalysed by cytochrome P450 is a monooxygenation, i.e., insertion of one atom of molecular oxygen into an organic substrate while the other oxygen atom is reduced to water:

$$RH + NADPH + H^{+}O_2 \rightarrow ROH + NADP^{+} + H_2O$$

The first step in the reaction cascade is an activation of the CYP-enzyme by a reduction that is catalysed by an NADPH dependent P450-reductase.

MFOs are responsible for the bioactivation of a wide variety of mostly nonpolar chemicals like PAHs and heterocyclic aromatic amines. Phase I reactions summable as oxidation, reduction, and hydrolysis mostly precede phase II reactions, but not

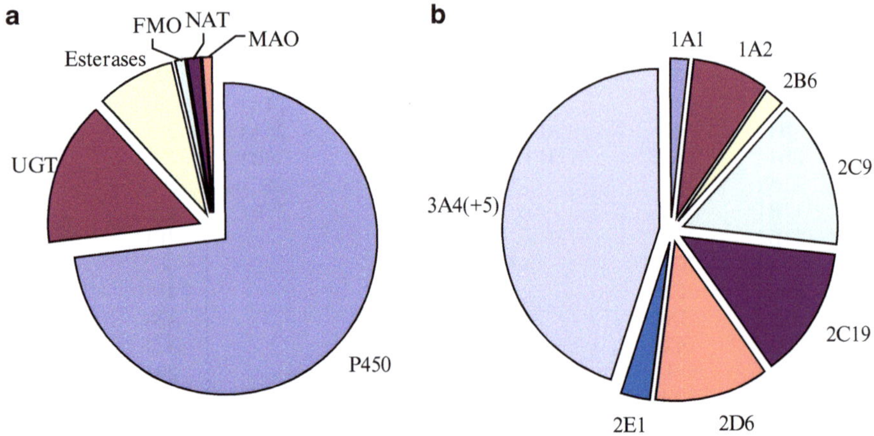

Fig. 5 Contribution of individual enzyme systems (**a**) and individual P450s (**b**) to the metabolism of xenobiotics (e.g., drugs). (Reprinted with permission from [81] Copyright 2006, Association of Pharmaceutical Scientists)

in all cases (Fig. 3). If the metabolites of phase I reactions are sufficiently polar, they may be readily excreted at this point; alternatively compounds might be converted directly by phase II reactions. However, many phase I products are not eliminated sufficiently and undergo a subsequent reaction in which an endogenous substrate is coupled to a newly synthesized functional group in order to form a highly polar conjugate.

Today an enormous number of bacterial P450s are known [82, 83]. More than 150 bacterial P450 families comprise more than 500 genes showing an extreme diversity. Compared to the eukaryotic P450s the protein structures are similar but bacterial MFOs are free-floating in the cell lacking the membrane anchoring part of their eukaryotic relatives that are membrane bound. *E. coli* has no P450 genes in its genome. Since bacterial cytochrome P450 genes do not play an important role in human genetic toxicology, the interest is focused on mammalian MFOs.

The first attempts to coexpress mammalian CYPs together with a P450-reductase in bacteria were made by Dong and Porter [84] who introduced P450 2E1 and rat NADPH-P450 reductase in *E. coli*. Low expression levels were overcome by using a two-promoter approach coexpressing P450 3A4 and human P450-reductase resulting in a high degree of catalytic activity in both, cells and isolated membranes [85]. The primary drawback in using living bacteria to catalyze mammalian P450-mediated reactions has been the paucity of electron transport from NADPH to P450 via endogenous flavoproteins. Shet et al. [86] successfully designed a bicistronic construct using a steroidogenic P450 (17A) and a P450 reductase. Parikh et al. [87] directly addressed the applicability of the system for the extension of the spectrum of P450 enzymes that are present in mammalian microsomes. They succeeded in functionally expressing bicistronic constructs in *E. coli* consisting out of human microsomal P450 enzymes and the auxiliary protein NADPH-P450-reductase encoded by the first and second cistron. They found that this approach has theoretical advantages compared to the coexpression systems by increasing the likelihood of colocalization of the two proteins within the bacterial membrane due to transcriptional/translational coupling in bacteria [87]. They got expression levels of P450s typically ranging from 100 to 200% of that of the NADPH-P450 reductase. In another approach, Emmert et al. [88] successfully expressed a complete electron transport chain consisting of P450 reductase, cytochrome B5 and cytochrome P450 2E1 in *S. typhimurium*. CYP 2E1 has an important role in human health as a result of being readily induced by acute and chronic alcohol ingestion. Typical substrates for 2E1 are low molecular chlorinated chemicals like allyl chloride, CCl_4, tri- and tetrachloroethylene and several nitroso derivatives. With the genetically engineered strain it was able to prove that several nitroso compounds and, e.g., propylene oxide and allyl chloride, that are negative in conventional tester strains, are activated to mutagenic metabolites by CYP 2E1 in the bacteria.

CYPs 1A1, 1A2, 1B1, 2E1, and 3A4 were identified to be the most relevant enzymes involved in the activation of procarcinogens and promutagens to DNA-reactive metabolites. In order to develop a very fast test system including intracellular phase I metabolic activity, Aryal and Oda et al. used several bicistronic CYP450/NADPH P450 reductase constructs in combination with the SOS/*umu* test principle [89, 90]. Efficient metabolic activation of heterocyclic amines like Glu-P-1, PhIp and Trp-P-1 were achieved by additionally introducing bacterial *O*-acetyltransferase (*O*-AT) [91]. Heterocyclic amines are formed during the cooking of proteinaceous food and have been identified to be highly carcinogenic [92]. They are widely distributed in a number of ambient environmental compartments such as airbone particles, diesel-exhaust particles, cigarette smoke, cooking fumes, rain water, sewage water, incineration ash and soil. These facts suggest that mutagenic heterocyclic amines are likely to be ubiquitous environmental pollutants and humans are continually exposed to these compounds in normal daily life [93].

8 Genetically Engineered Bacterial Strains with Phase-II Metabolic Competence

Not only microsomal cytochrome P450 enzymes but also several conjugative enzymes are known to play a key role in the (further) activation of reactive intermediates to genotoxic metabolites. Glutathione *S*-transferases, acetyltransferases (ATs) and sulfotransferases (SULTs) that are important in this context have been cloned in procaryotic cells (Tables 8 and 9).

In various animal species, including humans, there are multiple forms of glutathione *S*-transferase enzymes classified as alpha, pi, mu and theta. Theta-class GSTs are primarily involved in the detoxification of electrophilic metabolites of chemical carcinogens and mutagens [101, 102]. However, these enzymes also play an important role in the bioactivation of dihaloalkanes [103–106], compounds that have been widely used as pesticides, fumigants and antiknock substances [97].

Table 8 Genetically engineered strains for the detection of primary DNA-damage by means of the SOS-induction in bacteria with enhanced metabolic competence

Strain	Description	Specifity	Source or reference
S. typhimurium			
NM1011	Nitroreductase-overproducing strain	Nitroarenes	[94]
NM2009	*O*-AT-overexpressing strain[a]	Aromatic amines, hetero-cyclic aromatic amines	[95]
NM6001	As TA1538/1,8-DNP/pNM63/pSK1002, human NAT1-expressing strain	Aromatic amines, hetero-cyclic aromatic amines	[96]
NM6002	As TA1538/1,8-DNP/pNM63/pSK1002, human NAT2-expressing strain	Aromatic amines, hetero-cyclic aromatic amines	[96]
NM3009	Nitroreductase and ***O***-AT-overexpressing strain	Nitroarenes, aromatic amines	[121]
NM5004	GST-overexpressing strain (rat GST 5-5)	Dihaloalkanes	[97]

O-AT: *O*-acetyltransferase

Table 9 Genetically engineered strains for the detection of mutations in bacteria with enhanced metabolic competence

Strain	Description	Specifity	Source or reference
S. typhimurium			
YG1021	TA 98(pYG216) nitroreductase-overproducing	–2 Frameshifts, error prone repair, nitroarenes	[98]
TA98NR	TA 98, deficient in nitroreductase	–2 Frameshifts, error prone repair,	
YG1026	TA 100(pYG216) nitroreductase-overproducing	Base pair substitution, nitroarenes	[98]
TA100NR	TA 100, deficient in nitroreductase	Base pair substitution	
YG1024	*N*-Hydroxyarylamine *O*-AT-overexpressing strain	–2 Frameshifts, error prone repair, aromatic amines, aromatic amines	[99]
TA98/1,8-DNP6	TA 98, deficient in *O*-AT	–2 Frameshifts, error prone repair,	
YG1029	TA 100(pYG219) *O*-AT-overexpressing strain	Base pair substitution, error prone repair, aromatic amines, aromatic amines	[99]
TA100/1,8-DNP6	TA 100, deficient in *O*-AT	Base pair substitution	
YG1041	TA 98(pYG233) nitroreductase- and *O*-AT-overproducing	–2 Frameshifts, error prone repair, aromatic amines and nitroarenes	[100]
YG1042	TA 100(pYG233) nitroreductase- and *O*-AT-overproducing	Base pair substitution, error prone repair, aromatic amines and nitroarenes	[100]

Oda et al. [97] constructed a GST-overexpressing *umu* tester strain (NM5004) by using the theta-class rat GST 5-5, an enzyme that shares 82% sequence homology with human theta-class GST T1 [107]. The strain exhibits a 52-fold stronger GST activity than the parent strain TA1535/pSK1002 and was able to show the genotoxic potentials of ethylene dibromide, 1-bromo-2-chloroethane, 1,2-dichloroethane and methylene dichloride. With this study the participation of the SOS-response in the mutagenicity of dihaloalkanes has been proved.

AT and SULT enzymes are involved in the formation of reactive intermediates of arylamines, arylamides and PAHs [108]. Most of the aromatic amines are metabolized in two steps: *N*-oxidation by cytochrome P450 enzymes [109] and acetylation by *N,O*-acetyltransferase [110, 111] resulting in acetoxy esters, which form electrophilic, DNA reactive arylnitrenium ions. In 1990 Watanabe and colleagues developed a derivative of the Ames frameshift and base pair substitution tester strains TA98 and TA100 with high levels of *O*-AT activity [99]. Shortly thereafter, Yoshimitsu Oda provided a highly sensitive *umu* tester strain overexpressing bacterial AT for the detection of carcinogenic arylamines, aminoazo compounds and heterocyclic aromatic amines by the bacterial SOS-response [95]. Later on he cloned the human N-ATs (NAT1 and NAT2) via a plasmid vector into *S. typhimurium umu*-tester strains [96] in order to investigate the roles of human NAT enzymes in the genotoxicology of aromatic amines and nitroarenes. Human N-ATs are known to be polymorphic, expressing rapid, intermediate and slow acetylator phenotypes. Epidemiological studies have shown a relationship between urinary bladder cancer and slow acetylator phenotypes [112] and between colorectal cancer and rapid acetylator phenotypes [113].

Target cells of standard in vitro mutagenicity tests do not express any endogenous SULTs [114]. However, during phase II metabolism SULTs are often involved in the formation of DNA-reactive intermediates [115]. Aromatic amines, *N*-nitroso compounds and heterocyclic amines are suspected human pancreatic carcinogens. CYP 1A2, NAT1, NAT2 and SULT are enzymes involved in the metabolism of these carcinogens [116]. H. Glatt demonstrated for more than 100 chemicals a bioactivation by human SULTs to genotoxic metabolites directly after phase I metabolism [117]. For two reasons external activation of SULT substrates by liver homogenates is difficult: (1) external activation systems usually lack the SULT cofactor PAPS (3′-phosphoadenosine-5′ phosphosulfate) and (2) even after formation of the ultimate mutagen upon addition of the cofactor, sulfo-conjugated anions may not be able to enter the target cells. To solve these problems, Glatt and Meinl [118] developed methods to incorporate SULTs into in vitro mutagenicity test systems. They expressed most human SULTs in *S. typhimurium* TA 1538, TA 98 and TA 100 [119].

Beyond classical phase II enzymes, nitroreductases play a major role in the intracellular bioactivation of mutagenic and carcinogenic nitroarenes. Nitroarenes, a group of fused ring aromatic hydrocarbons which have one or more nitro moieties (e.g., 2-nitrofluorene, 1,8-dinitropyrene), are ubiquitous in the environment. They arise from incomplete combustion and oxides from nitrogen. Nitroarenes could be found in diesel and gasoline emission, cigarette smoke condensates, fly ash particles and in the urban atmosphere [119]. Nitro-containing chemicals get their DNA damaging potential by reduction of the nitro function to the corresponding hydroxylamine. However, the ultimate mutagens and carcinogens of aromatic azides, amino and nitro compounds are electrophilic nitrenium ions [120]. Several nitroreductase-overproducing tester strains have been cloned in recent years both for the Ames test [98] and for the SOS-tests [94]. The efficiency of detecting the mutagenicity of nitroarenes in the environment could again be enhanced by combining nitroreductase and AT activities in one tester strain [100, 121].

References

1. OECD (1997) Bacterial reverse mutation test. Guideline for Testing of Chemicals 471. OECD, Paris
2. Müller L, Kikuchi Y, Probst G, Schechtman L, Shimada H, Sofuni T, Tweats D (1999) ICH-Harmonised guidances on genotoxicity testing of pharmaceuticals: evolution, reasoning and impact. Mutat Res 436(3):195–225
3. OSPAR Commission (2002) Survey on genotoxicity test methods for the evaluation of waste water within whole effluent assessment. OSPAR, London, UK
4. ISO (2000) Water quality – determination of the genotoxicity of water and waste water using the *umu*-test. ISO 13829, ISO, Geneva
5. ISO (2005) Water quality – determination of the genotoxicity of water and waste water – *Salmonella*/microsome test (Ames test). ISO 16240, ISO, Geneva
6. International Conference on Harmonization (1997) Guidance for industry. S2B genotoxicity: a standard battery for genotoxicity testing of pharmaceuticals. Brussels, Belgium, March 1997
7. Hartman PE, Hartman Z, Stahl RC (1971) Classification and mapping of spontaneous and induced mutations in the histidine operon of *Salmonella*. Adv Genet 16:1–34
8. Ames BN, Gurney EG, Miller JA, Bartsch H (1972) Carcinogens as frameshift mutagens: metabolites and derivatives of 2-acetylaminofluorene and other aromatic amine carcinogens. Proc Natl Acad Sci U S A 69(11):3128–3132
9. Ames BN, Durston WE, Yamasaki E, Lee FD (1973) Carcinogens are mutagens: a simple test system combining liver homogenates for activation and bacteria for detection. Proc Natl Acad Sci U S A 70(8):2281–2285
10. McCann J, Ames BN (1976) Detection of carcinogens as mutagens in the *Salmonella*/microsome test: assay of 300 chemicals: discussion. Proc Natl Acad Sci U S A 73(3):950–954
11. Jurado J, Alejandre-Durán E, Pueyo C (1993) Genetic differences between the standard Ames tester strains TA100 and TA98. Mutagenesis. 8(6):527-32
12. McCann J, Spingarn NE, Kobori J, Ames BN (1975) Detection of carcinogens as mutagens: bacterial tester strains with R factor plasmids. Proc Natl Acad Sci U S A 72(3):979–983
13. Mortelmans K (2006) Isolation of plasmid pKM101 in the Stocker laboratory. Mutat Res 612(3):151–164
14. Cebula TA, Koch WH (1990) Sequence analysis of *Salmonella typhimurium* revertants. Prog Clin Biol Res. 340D:367–77. Review
15. Hartman PE, Ames BN, Roth JR, Barnes WM, Levin DE (1986) Target sequences for mutagenesis in *Salmonella* histidine-requiring mutants. Environ Mutagen 8(4):631–641
16. Walker GC (1977) Plasmid (pKM101)-mediated enhancement of repair and mutagenesis: dependence on chromosomal genes in *Escherichia coli* K-12. Mol Gen Genet 152(1):93–103
17. Marsh L, Walker GC (1987) New phenotypes associated with *mucAB*: alteration of a MucA sequence homologous to the LexA cleavage site. J Bacteriol 169(5):1818–1823
18. Perry KL, Elledge SJ, Mitchell BB, Marsh L, Walker GC (1985) *umuDC* and mucAB operons whose products are required for UV light- and chemical-induced mutagenesis: UmuD, MucA, and LexA proteins share homology. Proc Natl Acad Sci U S A 82(13):4331–4335

19. Levin DE, Hollstein M, Christman MF, Schwiers EA, Ames BN (1982) A new *Salmonella* tester strain (TA102) with A:T base pairs at the site of mutation detects oxidative mutagens. Proc Natl Acad Sci U S A 79(23):7445–7449
20. Albertini S, Gocke E (1988) Plasmid copy number and mutant frequencies in *S. typhimurium TA102*. Environ Mol Mutagen 12(4):353–363
21. Levin DE, Yamasaki E, Ames BN (1982) A new *Salmonella* tester strain, TA97, for the detection of frameshift mutagens. A run of cytosines as a mutational hot-spot. Mutat Res 94(2): 315–330
22. Bridges BA, Dennis RE, Munson RJ (1967) Differential induction and repair of ultraviolet damage leading to true revesions and external suppressor mutations of an ochre codon in *Escherichia coli* B-r WP2. Genetics 57(4):897–908
23. Venturini S, Monti-Bragadin C (1978) R plasmid-mediated enhancement of mutagenesis in strains of Escherichia coli deficient is known repair functions. Mutat Res 50(1):1–8
24. Prival MJ, King VD, Sheldon AT Jr (1979) The mutagenicity of dialkyl nitrosamines in the *Salmonella plate* assay. Environ Mutagen 1(2):95–104
25. Agurell E, Stensman C (1992) Salmonella mutagenicity of three complex mixtures assayed with the microsuspension technique. A WHO/IPCS/CSCM study. Mutat Res 276(1/2): 87–91
26. Bagley ST, Stoltz SL, Becker DM, Keen RE (1992) Characterization of organic extracts from standard reference materials 1649, 'urban dust/organics,' and 1650, 'diesel particulate matter', using a microsuspension assay. A WHO/IPCS/CSCM study. Mutat Res 276(1/2):81–86
27. Azuma S, Kishino S, Katayama S, Akahori Y, Matsushita H (1997) Highly sensitive mutation assay for mutagenicity monitoring of indoor air using *Salmonella typhimurium* YG1041 and a microsuspension method. Mutagenesis 12(5):373–377
28. Gatehouse DG, Paes DJ (1983) A demonstration of the in vitro bacterial mutagenicity of procarbazine, using the microtitre fluctuation test and large concentrations of S9 fraction. Carcinogenesis 4(3):347–352
29. Flückiger-Isler S, Baumeister M, Braun K, Gervais V, Hasler-Nguyen N, Reimann R, Van Gompel J, Wunderlich HG, Engelhardt G (2004) Assessment of the performance of the Ames II assay: a collaborative study with 19 coded compounds. Mutat Res 558(1/2):181–197
30. Reifferscheid G, Arndt C, Schmid C (2005) Further development of the beta-lactamase MutaGen assay and evaluation by comparison with Ames fluctuation tests and the *umu* test. Environ Mol Mutagen 46(2):126–139
31. Gee P, Maron DM, Ames BN (1994) Detection and classification of mutagens: a set of base-specific *Salmonella* tester strains. Proc Natl Acad Sci U S A 91(24):11606–11610
32. Cupples CG, Miller JH (1989) A set of *lacZ* mutations in *Escherichia coli* that allow rapid detection of each of the six base substitutions. Proc Natl Acad Sci U S A 86(14):5345–5349
33. Miller JE, Vlasakova K, Glaab WE, Skopek TR (2005) A low volume, high-throughput forward mutation assay in *Salmonella typhimurium* based on fluorouracil resistance. Mutat Res 578(1/2):210–224
34. Dorado G, Pueyo C (1988) L-Arabinose resistance test with *Salmonella typhimurium* as a primary tool for carcinogen screening. Cancer Res 48(4):907–912
35. Schmid C, Arndt C, Reifferscheid G (2003) Mutagenicity test system based on a reporter gene assay for short-term detection of mutagens (MutaGen assay). Mutat Res 535(1):55–72
36. Tomasz A (1979) From penicillin-binding proteins to the lysis and death of bacteria: a 1979 view. Rev Infect Dis 1(3):434–467
37. Tomasz A (1979) The mechanism of the irreversible antimicrobial effects of penicillins: how the beta-lactam antibiotics kill and lyse bacteria. Annu Rev Microbiol 33:113–137
38. Hillen W, Berens C (1994) Mechanisms underlying expression of TN10 encoded tetracycline resistance. Annu Rev Microbiol 48:345–369
39. Degenkolb J, Takahashi M, Ellestad GA, Hillen W (1991) Structural requirements of tetracycline-*tet* repressor interaction: determination of equilibrium binding constants for tetracycline analogs with the tet repressor. Antimicrob Agents Chemother 35:1591–1595
40. Radman M (1975) SOS repair hypothesis: phenomenology of an inducible DNA repair which is accompanied by mutagenesis. Basic Life Sci 5A:355–367

41. Brent R, Ptashne M (1980) The *lexA* gene product represses its own promoter. Proc Natl Acad Sci U S A 77(4):1932–1936
42. Little JW, Mount DW, Yanisch-Perron CR (1981) Purified lexA protein is a repressor of the *recA* and *lexA* genes. Proc Natl Acad Sci U S A 78(7):4199–4203
43. Little JW (1983) The SOS regulatory system: control of its state by the level of RecA protease. J Mol Biol 167(4):791–808
44. Little JW, Edmiston SH, Pacelli LZ, Mount DW (1980) Cleavage of the *Escherichia coli* lexA protein by the recA protease. Proc Natl Acad Sci U S A 77(6):3225–3229
45. Giese KC, Michalowski CB, Little JW (2008) RecA-dependent cleavage of LexA dimers. J Mol Biol 377(1):148–161
46. Huisman O, D'Ari R, George J (1980) Further characterization of *sfiA* and *sfiB* mutations in *Escherichia coli*. J Bacteriol 144(1):185–191
47. Schendel PF, Fogliano M, Strausbaugh LD (1982) Regulation of the *Escherichia coli* K-12 uvrB operon. J Bacteriol 150(2):676–685
48. Frey J, Ghersa P, Palacios PG, Belet M (1986) Physical and genetic analysis of the ColD plasmid. J Bacteriol. 166(1):15–9
49. Tomita K, Ogawa T, Uozumi T, Watanabe K, Masaki H (2000) A cytotoxic ribonuclease which specifically cleaves four isoaccepting arginine tRNAs at their anticodon loops. Proc Natl Acad Sci U S A 97(15):8278–8283
50. de Zamaroczy M, Mora L, Lecuyer A, Géli V, Buckingham RH (2001) Cleavage of colicin D is necessary for cell killing and requires the inner membrane peptidase LepB. Mol Cell 8(1):159–168
51. Quillardet P, Huisman O, D'Ari R, Hofnung M (1982) SOS chromotest, a direct assay of induction of an SOS function in *Escherichia coli* K-12 to measure genotoxicity. Proc Natl Acad Sci U S A 79(19):5971–5975
52. Oda Y, Nakamura S, Oki I, Kato T, Shinagawa H (1985) Evaluation of the new system (*umu*-test) for the detection of environmental mutagens and carcinogens. Mutat Res 147(5):219–229
53. Ptitsyn LR, Horneck G, Komova O, Kozubek S, Krasavin EA, Bonev M, Rettberg P (1997) A biosensor for environmental genotoxin screening based on an SOS lux assay in recombinant *Escherichia coli* cells. Appl Environ Microbiol. 63(11):4377–84
54. Norman A, Hestbjerg Hansen L, Sørensen SJ (2005) Construction of a ColD *cda* promoter-based SOS-green fluorescent protein whole-cell biosensor with higher sensitivity toward genotoxic compounds than constructs based on *recA*, *umuDC*, or sulA promoters. Appl Environ Microbiol 71(5):2338–2346
55. Nunoshiba T, Nishioka H (1991) 'Rec-lac test' for detecting SOS-inducing activity of environmental genotoxic substance. Mutat Res 254(1):71–77
56. Verschaeve L, Van Gompel J, Thilemans L, Regniers L, Vanparys P, van der Lelie D (1999) VITOTOX bacterial genotoxicity and toxicity test for the rapid screening of chemicals. Environ Mol Mutagen 33(3):240–248
57. Vollmer AC, Belkin S, Smulski DR, Van Dyk TK, LaRossa RA (1997) Detection of DNA damage by use of *Escherichia coli* carrying *recA'::lux*, *uvrA'::lux* or *alkA'::lux* reporter plasmids. Appl Environ Microbiol 63:2566–2571
58. Davidov Y, Rozen R, Smulski DR, van Dyk TK, Vollmer AC, Elsemore DA, LaRossa RA, Belkin S (2000) Improved bacterial SOS promoter: *lux* fusions for genotoxicity detection. Mutat Res 466:97–107
59. Chalfie M, Tu G, Euskirchen G, Ward WW, Prasher DC (1994) Green fluorescent protein as a marker for gene expression. Science 263:802–804
60. Crameri A, Whitehorn EA, Tate E, Stemmer WPC (1996) Improved green fluorescent protein by molecular evolution using DNA shuffling. Nat Biotechnol 14:315–319
61. Arkady FF, Ying C, Li D, Ekaterina VB, Mikhail VM, Sergey AL (2000) Novel fluorescent protein from Discosoma coral and its mutants possesses a unique far-red fluorescence. FEBS Lett 479:127–130
62. Shaner NC, Patterson GH, Davidson MW (2007) Advances in fluorescent protein technology. J Cell Sci 120:4247–4260

63. Nagai T, Ibata K, Park ES, Kubota M, Mikoshiba K, Miyawaki A (2002) A variant of yellow fluorescent protein with fast and efficient maturation for cell-biological applications. Nat Biotechnol 20:87–90
64. Blokpoel MC, O'Toole R, Smeulders MJ, Williams HD (2003) Development and application of unstable GFP variants to kinetic studies of mycobacterial gene expression. J Microbiol Methods 54:203–211
65. Justus T, Thomas SM (1999) Evaluation of transcriptional fusions with green fluorescent protein versus luciferase as reporters in bacterial mutagenicity tests. Mutagenesis 14:351–356
66. Arai R, Makita Y, Oda Y, Nagamune T (2001) Construction of green fluorescent protein reporter genes for genotoxicity test (SOS/*umu*-test) and improvement of mutagen-sensitivity. J Biosci Bioeng 92:301–304
67. Kostrzynska M, Leung KT, Lee H, Trevors JT (2002) Green fluorescent protein-based biosensor for detecting SOS-inducing activity of genotoxic compounds. J Microbiol Methods 48:43–51
68. Sagi E, Hever N, Rosen R, Bartolome AJ, Rajan Premkumar J, Ulber R, Lev O, Scheper T, Belkin S (2003) Fluorescence and bioluminescence reporter functions in genetically modified bacterial sensor strains. Sens Actuators B Chem 90:2–8
69. Hakkila K, Maksimow M, Karp M, Virta M (2002) Reporter genes *lucFF*, *luxCDABE*, *gfp*, and *dsred* have different characteristics in whole-cell bacterial sensors. Anal Biochem 301:235–242
70. Norman A, Hansen LH, Sorensen SJ (2006) A flow cytometry-optimized assay using an SOS-green fluorescent protein (SOS-GFP) whole-cell biosensor for the detection of genotoxins in complex environments. Mutat Res 603(2):164–172
71. Ostergaard TG, Hansen LH, Binderup ML, Norman A, Sørensen SJ (2007) The *cda* GenoTox assay: a new and sensitive method for detection of environmental genotoxins, including nitroarenes and aromatic amines. Mutat Res 631(2):77–84
72. von der Hude W, Behm C, Gürtler R, Basler A (1988) Evaluation of the SOS chromotest. Mutat Res 203(2):81–94
73. Dipple A (1995) DNA adducts of chemical carcinogens. Carcinogenesis 16(3):437–441
74. Seidegard J, Ekstrom G (1997) The role of human glutathione transferases and epoxide hydrolases in the metabolism of xenobiotics. Environ Health Perspect 105(4):791–799
75. Kranendonk M, Laires A, Rueff J, Estabrook WR, Vermeulen NP (2000) Heterologous expression of xenobiotic mammalian-metabolizing enzymes in mutagenicity tester bacteria: an update and practical considerations. Crit Rev Toxicol 30:287–306
76. Shimada T, Yamazaki H, Oda Y, Hiratsuka A, Watabe T, Guengerich FP (1996) Activation and inactivation of carcinogenic dihaloalkanes and other compounds by glutathione *S*-transferase 5-5 in *Salmonella typhimurium* tester strain NM5004. Chem Res Toxicol 9:333–340
77. Martignoni M, Groothuis GM, de Kanter R (2006) Species differences between mouse, rat, dog, monkey and human CYP-mediated drug metabolism, inhibition and induction. Expert Opin Drug Metab Toxicol 2(6):875–894
78. Gundert-Remy U, Sonich-Mullin C (2002) IPCS uncertainty and variability planning workgroup and drafting group (International program on chemical safetys). The use of toxicokinetic and toxicodynamic data in risk assessment: an international perspective. Sci Total Environ 288(1/2):3–11
79. Malling HV (1967) The mutagenicity of the acridine mustard (ICR-170) and the structurally related compounds in Neurospora. Mutat Res 4(3):265–274
80. Glatt H (1997) Sulfation and sulfotransferases 4: bioactivation of mutagens via sulfation. FASEB J 11(5):314–321
81. Guengerich FP (2006) Cytochrome P450s and other enzymes in drug metabolism and toxicity. AAPS J 8(1):101–111
82. Roberts GA, Grogan G, Greter A, Flitsch SL, Turner NJ (2002) Identification of a new class of cytochrome P450 from a *Rhodococcus sp*. J Bacteriol 184(14):3898–3908
83. Agematu H, Matsumoto N, Fujii Y, Kabumoto H, Doi S, Machida K, Ishikawa J, Arisawa A (2006) Hydroxylation of testosterone by bacterial cytochromes P450 using the *Escherichia coli* expression system. Biosci Biotechnol Biochem 70(1):307–311
84. Dong J, Porter TD (1996) Coexpression of mammalian cytochrome P450 and reductase in *Escherichia coli*. Arch Biochem Biophys 327(2):254–259

85. Blake JA, Pritchard M, Ding S, Smith GC, Burchell B, Wolf CR, Friedberg T (1996) Coexpression of a human P450 (CYP3A4) and P450 reductase generates a highly functional monooxygenase system in *Escherichia coli*. FEBS Lett 397(2/3):210–214
86. Shet MS, Fisher CW, Estabrook RW (1997) The function of recombinant cytochrome P450s in intact *Escherichia coli* cells: the 17 alpha-hydroxylation of progesterone and pregnenolone by P450c17. Arch Biochem Biophys 339(1):218–225
87. Parikh A, Gillam EM, Guengerich FP (1997) Drug metabolism by *Escherichia coli* expressing human cytochromes P450. Nat Biotechnol 15(8):784–788
88. Emmert B, Bünger J, Keuch K, Müller M, Emmert S, Hallier E, Westphal GA (2006) Mutagenicity of cytochrome P450 2E1 substrates in the Ames test with the metabolic competent *S. typhimurium* strain YG7108pin3ERb5. Toxicology 228(1):66–76
89. Aryal P, Yoshikawa K, Terashita T, Guengerich FP, Shimada T, Oda Y (1999) Development of a new genotoxicity test system with *Salmonella typhimurium* OY1001/1A2 expressing human CYP1A2 and NADPH-P450 reductase. Mutat Res 442(2):113–120
90. Oda Y, Aryal P, Terashita T, Gillam EM, Guengerich FP, Shimada T (2001) Metabolic activation of heterocyclic amines and other procarcinogens in *Salmonella typhimurium umu* tester strains expressing human cytochrome P4501A1, 1A2, 1B1, 2C9, 2D6, 2E1, and 3A4 and human NADPH-P450 reductase and bacterial O-acetyltransferase. Mutat Res 492(1/2):81–90
91. Aryal P, Terashita T, Guengerich FP, Shimada T, Oda Y (2000) Use of genetically engineered *Salmonella typhimurium* OY1002/1A2 strain coexpressing human cytochrome P450 1A2 and NADPH-cytochrome P450 reductase and bacterial *O*-acetyltransferase in SOS/*umu* assay. Environ Mol Mutagen 36(2):121–126
92. Sugimura T (1997) Overview of carcinogenic heterocyclic amines. Mutat Res 376(1/2):211–219
93. Kataoka H (1997) Methods for the determination of mutagenic heterocyclic amines and their applications in environmental analysis. J Chromatogr A 774(1/2):121–142
94. Oda Y, Shimada T, Watanabe M, Ishidate M Jr, Nohmi T (1992) A sensitive *umu* test system for the detection of mutagenic nitroarenes in *Salmonella typhimurium* NM1011 having a high nitroreductase activity. Mutat Res 272(2):91–99
95. Oda Y, Yamazaki H, Watanabe M, Nohmi T, Shimada T (1995) Development of high sensitive umu test system: rapid detection of genotoxicity of promutagenic aromatic amines by *Salmonella typhimurium* strain NM2009 possessing high O-acetyltransferase activity. Mutat Res 334(2):145–156
96. Oda Y, Yamazaki H, Shimada T (1999) Role of human *N*-acetyltransferases, NAT1 or NAT2, in genotoxicity of nitroarenes and aromatic amines in *Salmonella typhimurium* NM6001 and NM6002. Carcinogenesis 20(6):1079–1083
97. Oda Y, Yamazaki H, Thier R, Ketterer B, Guengerich FP, Shimada T (1996) A new *Salmonella typhimurium* NM5004 strain expressing rat glutathione S-transferase 5-5: use in detection of genotoxicity of dihaloalkanes using an SOS/*umu* test system. Carcinogenesis 17(2):297–302
98. Watanabe M, Ishidate M Jr, Nohmi T (1989) A sensitive method for the detection of mutagenic nitroarenes: construction of nitroreductase-overproducing derivatives of *Salmonella typhimurium* strains TA98 and TA100. Mutat Res 216(4):211–220
99. Watanabe M, Ishidate M Jr, Nohmi T (1990) Sensitive method for the detection of mutagenic nitroarenes and aromatic amines: new derivatives of *Salmonella typhimurium* tester strains possessing elevated O-acetyltransferase levels. Mutat Res 234(5):337–348
100. Hagiwara Y, Watanabe M, Oda Y, Sofuni T, Nohmi T (1993) Specificity and sensitivity of *Salmonella typhimurium* YG1041 and YG1042 strains possessing elevated levels of both nitroreductase and acetyltransferase activity. Mutat Res 291(3):171–180
101. Chasseaud LF (1997) The role of glutathione ans glutathione transferases in the metabolism of chemical carcinogens and other electrophilic reagents. Adv Cancer Res 29:175–274
102. Coles B, Ketterer B (1990) The role of glutathione and glutathione transfersase in chemical carcinogenesis. Crit Rev Biochem Mol Biol 25:47–70.

103. Decant W, Vamvakas S, Anders MW (1989) Bioactivation of nephrotoxic haloalkenes by glutathiuone conjugation: formation of toxic and mutagenic intermediates by cysteine conjugate ß-lyase. Drug Metab Rev 20:43–83
104. Monks TJ, Lau SS (1994) Glutathione conjugate mediated toxicities. In: Kauffman FC (ed) Handbook of experimental pharmacology. Springer, Berlin Heidelberg New York, pp 459–509
105. Thier R, Taylor JB, Pemble SE, Humphreys WG, Persmark M, Ketterer B, Guengerich FP (1993) Expression of mammalian glutathione S-transferase 5-5 in *Salmonella typhimurium* TA1535 leads to base-pair mutations upon exposure to dihalomethanes. Proc Natl Acad Sci U S A 90(18):8576–8580
106. Guengerich FP, Wheeler JB, Chun YJ, Kim D, Shimada T, Aryal P, Oda Y, Gillam EM (2002) Use of beterologously-expressed cytochrome P450 and glutathione transferase enzymes in toxicity assays. Toxicology 181–182:261–4
107. Pemble S, Schroeder KR, Spencer SR, Meyer DJ, Hallier E, Bolt HM, Ketterer B, Taylor JB (1994) Human glutathione S-transferase theta (GSTT1): cDNA cloning and the characterization of a genetic polymorphism. Biochem J 300(1):271–276
108. Yamazaki H, Oda Y, Shimada T (1992) Use of a newly developed tester strain *Salmonella typhimurium* NM2009 for the study of metabolic activation of carcinogenic aromatic amines by rat liver microsomal cytochrome P-450 enzymes. Mutat Res 272(2):183–192
109. Lotlikar PD, Hong YS (1981) Microsomal N- and C-oxidations of carcinogenic aromatic amines and amides. Natl Cancer Inst Monogr 58:101–107
110. McCoy EC, Anders M, Rosenkranz HS (1983) The basis of the insensitivity of *Salmonella typhimurium* strain TA98/1,8-DNP6 to the mutagenic action of nitroarenes. Mutat Res 121(1):17–23
111. Heflich RH, Djuri Z, Zhuo Z, Fullerton NF, Casciano DA, Beland FA (1988) Metabolism of 2-acetylaminofluorene in the Chinese hamster ovary cell mutation assay. Environ Mol Mutagen 11(2):167–181
112. Lower GM Jr, Nilsson T, Nelson CE, Wolf H, Gamsky TE, Bryan GT (1979) N-Acetyltransferase phenotype and risk in urinary bladder cancer: approaches in molecular epidemiology. Preliminary results in Sweden and Denmark. Environ Health Perspect 29:71–79
113. Lang NP, Chu DZ, Hunter CF, Kendall DC, Flammang TJ, Kadlubar FF (1986) Role of aromatic amine acetyltransferase in human colorectal cancer. Arch Surg 121(11):1259–1261
114. Glatt H, Meinl W (2005) Sulfotransferases and acetyltransferases in mutagenicity testing: technical aspects. Methods Enzymol 400:230–249
115. Miller JA, Surh YJ (1994) Sulfonation in chemical carcinogenes. In: Kauffman FC (ed) Handbook of pharmacology, vol 112: Conjugation-deconjugation reactions in drug metabolism and toxicity. Springer, Berlin Heidelberg New York, pp 429–458
116. Suzuki H, Morris JS, Li Y, Doll MA, Hein DW, Liu J, Jiao L, Hassan MM, Day RS, Bondy ML, Abbruzzese JL, Li D (2008) Interaction of the cytochrome P4501A2, SULT1A1 and NAT gene polymorphisms with smoking and dietary mutagen intake in modification of the risk of pancreatic cancer. Carcinogenesis 29(6):1184–1191
117. Glatt H (2005) Activation and inactivation of carcinogens by human sulfotransferases. In: Pacifici GM, Coughtrie MWH (eds) Human sulfotransferases. Taylor & Francis, London
118. Glatt H, Meinl W (2004) Use of genetically manipulated *Salmonella typhimurium* strains to evaluate the role of sulfotransferases and acetyltransferases in nitrofen mutagenicity. Carcinogenesis 25(5):779–786
119. Tokiwa H, Ohnishi Y (1986) Mutagenicity and carcinogenicity of nitroarenes and their sources in the environment. Crit Rev Toxicol 17(1):23–60
120. Wild D (1990) A novel pathway to the ultimate mutagens of aromatic amino and nitro compounds. Environ Health Perspect 88:27–31
121. Oda Y, Yamazaki H, Watanabe M, Nohmi T, Shimada T (1993) Highly sensitive *umu* test system for the detection of mutagenic nitroarenes in *Salmonella typhimurium* NM3009 having high *O*-acetyltransferase and nitroreductase activities. Environ Mol Mutagen 21(4):357–364

Adv Biochem Engin/Biotechnol (2010) 118: 113–151
DOI: 10.1007/10_2009_20

Published online: 6 February 2010

Cytotoxicity and Genotoxicity Reporter Systems Based on the Use of Mammalian Cells

Christa Baumstark-Khan, Christine E. Hellweg, and Günther Reitz

Abstract With the dramatic increase in the number of new agents arising from the chemical, pharmaceutical, and agricultural industries, there is an urgent need to develop assays for rapid evaluation of potential risks to man and environment. The panel of conventional tests used for cytotoxicity and genotoxicity and the strategies to progress from small scale assays to high content screening in toxicology are discussed. The properties of components necessary as sensors and reporters for new reporter assays, and the application of genetic strategies to design assays are reviewed. The concept of cellular reporters is based on the use of promoters of chemical stress-regulated genes ligated to a suitable luminescent or fluorescent reporter gene. Current reporter assays designed from constructs transferred into suitable cell lines are presented.

Keywords Cytotoxicity • DNA damage and repair • Genotoxicity • Mammalian cells • Reporter assays

Contents

C. Baumstark-Khan (✉), C.E. Hellweg, and G. Reitz
Radiation Biology Department, Institute of Aerospace Medicine, German Aerospace Centre (DLR), Linder Hoehe, 51147 Cologne, Germany
e-mail: christa.baumstark-khan@dlr.de

1 Introduction

Development of new products for agricultural, pharmaceutical, and consumer use is a difficult, time consuming, and expensive process. This process often takes more than 10 years and costs hundreds of millions of dollars. Generally, several thousands of promising molecules have to be examined for efficacy and for safety to find a single product for marketing. The new European Community Regulation on chemicals and their safe use REACH (EC 1907/2006) deals with the registration, evaluation, authorization, and restriction of chemical substances [1]. The new law entered into force on 1 June 2007 and aims to improve the protection of human health and the environment through better and earlier identification of the intrinsic properties of chemical substances. Toxicology studies play a central role in (1) evaluating health hazards associated with the exposure of humans to these products, (2) making crucial decisions on whether or not to invest valuable resources in developing a new lead molecule, and (3) to maintain a product that is already on the market by providing data from up-to-date state of the art study designs [2, 3].

The method for the determination of acute toxicity was the LD_{50} Acute Toxicity Test, introduced by Trevan in 1927 [4] for biological standardization of potent and potentially dangerous drugs such as digitalis extracts, insulin and diphtheria toxin. It was a standardized measure for expressing and comparing the toxicity of chemicals. The median lethal dose, LD_{50} (abbreviation for "Lethal Dose, 50%"), or LCt_{50} (Lethal Concentration and Time) of a toxic substance or radiation is the dose that kills half [5] of the animals tested (LD = "lethal dose"). The animals are usually rats or mice, although rabbits, guinea pigs, hamsters, and so on are sometimes used. The mode of chemical application may vary from oral to intravenous administration. The choice of 50% lethality as a benchmark avoids the potential for ambiguity of making measurements at the extremes, and reduces the amount of testing required. However, this also means that LD_{50} is not the lethal dose for all subjects; some may be killed by much less, while others survive doses far higher than the LD_{50}. Results of LD_{50} tests have been used to predict lethal doses and symptoms of human intoxications. Since LD_{50} values may vary greatly among different species, prediction of human lethal doses from rodent LD_{50} should be made with caution. In the past three decades there has been a move away from the classical LD_{50} test, which used up to 100 animals, towards alternative methods that use fewer animals (6–10) and reduce the trauma to them and since 2002 an agreement has been reached on the abolition of the LD_{50} Acute Toxicity Test [6]. A method that uses lower doses and focuses on signs of toxicity rather than death is the Fixed Dose Procedure [7]. The Acute Toxic Class Method [8] and the Up-and-Down Procedure [9] use fewer animals, but death is still the endpoint.

Generally, acute toxicity tests in animals are not particularly useful for hazard identification and risk assessment in relation to foods and environmental chemicals because human exposure tends to be much lower than that which causes acute toxicity and continues over much longer periods. Thus, for human populations the major concern is the induction and promotion of tumors as a result of environmental and anthropogenic exposure. The "gold standard" test for determining the carcinogenic activity of a test chemical is the chronic rodent carcinogenicity bioassay [10, 11]. Though expensive, the bioassay is widely conducted and is prescribed by regulatory authorities for many products, such as drugs, food additives, and pesticides.

A number of issues on the use of animals in toxicity research have arisen in the scientific, regulatory and public communities over the past several years. There are consistently increasing concerns on the rights of the animals used for research and on the already high and ever increasing costs of tests performed with a limited number of animals [12, 13]. Accordingly, the chemical and cosmetics industries have demonstrated an increasing interest in the development of in vitro methods that have the potencies for refinement, reduction, and replacement of conventional animal tests [14–16]. This can be achieved by methods which allow for better understanding of targets and toxicity mechanisms, by methods which introduce screening methods for preselecting the compounds of highest concern and which show up by better efficiencies.

2 Conventional In Vitro Assays for Assessing Cytotoxicity

From the perspective of pharmaceutical drug development, cytotoxicological studies are intended to (1) identify an initial safe starting dose and subsequent dose escalation scheme to humans, (2) potential target organs of toxicity and reversibility of toxicity, and (3) parameters for clinical monitoring [17].

Most of the current in vitro cytotoxicity testing procedures measure a negative signal, i.e., cell death. This means that a certain result, lethality, is measured after irreversible damage has occurred. Despite the variety of methods currently available for measurement of cellular damage, the question of which assay provides the best estimate of viability remains. The general opinion is that the most consistent, relevant, and reproducible answer is given by the in vitro colony formation assay which was developed for experimentation with ionizing radiation [18, 19]. Testing colony formation ability as a measure for cell survival requires investigation of the reproductive potential of proliferating cells in culture to a chemical or physical damage. When used on adherent cells, the test involves obtaining a single cell suspension either by enzymatic or mechanical detachment of the desired cell type from the surface of the growth vessel, followed by counting. Known numbers of cells are plated onto Petri dishes at low cell to surface area ratios for growth and division (Fig. 1). Under such conditions, viable cells will divide and eventually give rise to a colony or cluster of daughter cells, the number of which depends on initial

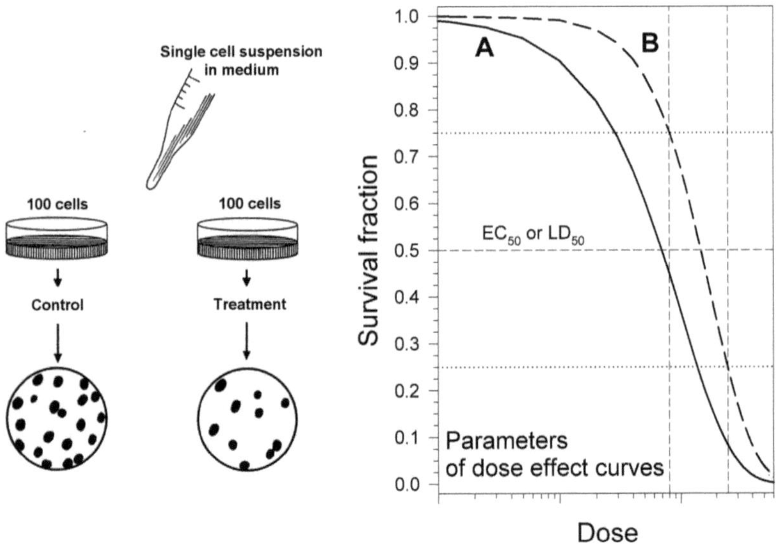

Fig. 1 Performance and outcome of the colony forming ability test (CFA)

plating efficiency of the cell line of question and any anticipated effect resulting from incubation in the presence of a cytotoxic agent. Generally, cells are considered as viable if they are capable of dividing at least seven times, i.e., giving rise to a colony of greater than 50 cells. The incubation time necessary to give rise to countable colonies depends on the population growth rate. It can vary from 5 to 7 days for a cell line such as the Chinese hamster lung fibroblast (V79) with a 9-h doubling time whereas 2–3 weeks of incubation may be required for a cell line with a 24-h doubling time such as human fibroblasts or even tumor cells.

The CFA assay is time consuming and labor intensive, particularly when many samples have to be processed. Accordingly, other assays are being searched for, which supply reliable data within a reasonable time. One alternative to the CFA assay is measuring growth properties of cell cultures, especially measuring the absolute increase in cell numbers over an extended period of time, generally by recording the increase in the number of cells at timed intervals. Kinetic characteristics of undisturbed cultures, which depend on the cell type used, include an initial lag phase after plating, followed by an exponential increase in cell numbers, and ending with a plateau region either when no further surface area is available for growth or when vital nutrients are depleted. Growth survival values are obtained by plotting cell numbers against time and extrapolating from the linear portion of the growth curve back to the Y-intercept. This derived cell number can then be compared to the calculated cell number of an untreated population.

Besides assays testing for cellular reproducibility as a measure for toxicity, a series of functional assays has been developed. Functional assays typically evaluate viability by examining metabolic components that are necessary for growth. Metabolism, as an essential process for the survival of organisms, serves two fundamentally important functions in the cells: generation of biological energy to drive vital functions and the formation of precursors, which are essential for the synthesis of biological molecules and cellular constituents.

The basis of such tests is the premise that cellular damage will inevitably result in loss of the ability to maintain and provide energy for metabolic function and growth. Such effects can be analyzed by several methods, e.g., counting cells that include/exclude a dye, measuring released ^{51}Cr-labeled protein after cell lysis [20], and measuring incorporation of radioactive nucleotides such as ^{3}H-thymidine [21] or ^{125}I-iodo deoxyuridine [22] during cell proliferation. The radioactive methods can be partially automated and can handle moderately large numbers of samples. However, a lot of radioactive waste is produced which creates huge costs, as it needs to be stored for long time periods to decay. Accordingly, nonradioactive methods are preferred nowadays.

The ATP assay is a short term in vitro cytotoxicity assay based on metabolic activity measured as cellular ATP content [23, 24]. The amount of ATP in a specific cell type is relatively constant [25, 26]. ATP is rapidly degraded by ATPases leading to prompt depletion if the respiratory cycle is disturbed in aerobic cells. The intracellular ATP is released by TCA and the ATP levels can be measured directly from an aliquot of the growth medium without any extraction or precipitation steps. Since the ATP levels are constant in a given healthy cell, it can be used as an indirect method for measuring cell growth or death [27]. A variant of the ATP cell viability assay is the microplate ATP-based tumor chemo sensitivity assay (ATP-TCA), which overcame major technical limitations of older assays such as low degree of standardization and reproducibility, lack of technical robustness, and poor methodological efficacy. It can now be considered the best documented and validated technology [28]. This assay, which is now commercially available, provides a highly reproducible, easy-to-handle kit technique, low technical failure rates, and a high methodological efficacy. In primary ovarian cancers, the ATP-TCA has been found to predict accurately both therapy response and patient survival as a milestone for individualized tumor therapy.

There are several others commonly used cytotoxicity screening assays that are based on basic functions of living cells. The MTT-based assay [29] determines the ability of viable cells with active mitochondria to convert the soluble tetrazolium salt 3-(4,5-dimethylthiazol-2-yl)-2,5-diphenyltetrazolium bromide (MTT) into an insoluble dark blue formazan precipitate, the amount of which depends on the level of cellular energy metabolism. The test is easy to perform and is adaptable for high throughput of samples, although there are some limitations in its application resulting from metabolic interference [29–31]. The neutral red uptake test [32, 33] and the LDH release test [34] make use of membrane barrier functions of healthy cells. Another parameter related to cellular metabolism and capable of showing cytotoxicity is macromolecular synthesis [35], especially the synthesis of DNA. However, an increase in macromolecular contents is not necessarily correlated to cell multiplication, as changes in cell volumes in nondividing populations are reflected [36]. The RoDos dye binding assay [37] was especially designed for determination of cytotoxicity induced by UV and visible light. It is based on rodent cells growing on an UV-transparent Lumox foil and a special device which allows irradiation with different doses of UV-light, assuring cellular growth under identical spatial arrangement of the cells. The influence of irradiation on "cell growth" is determined by image analysis as the amount of the bound dye crystal violet of the irradiated to unirradiated areas of the Lumox dish.

Over recent years, the concept of cell death has changed radically as a result of the widespread interest in apoptosis, the process of programmed cell death. This has lead to an emerging understanding how control mechanisms of apoptosis break down in diseases [38]. The socio-economic importance of the degenerative diseases is fuelling considerable research effort into cell death in both academia and the pharmaceutical industry. Of great importance to the study of cell death is the differentiation between the two principal modes, namely, apoptosis and necrosis [39]. Apoptosis is an active process that is accompanied by a series of distinct cellular and molecular events that form an integral part of normal physiological processes. These include reduction in cell size, condensation of nuclear chromatin, and activation of endogenous endonucleases and proteases, which induce intracellular digestion of the cell. In vivo, apoptosis is followed by removal of the cell by phagocytosis while necrosis is the passive result of cellular injury leading to complete loss of cell integrity and the release of cellular components. Necrosis results in a wide range of undesirable side effects including inflammatory responses. Flow cytometric analysis remains the method of choice for the study of apoptosis and necrosis[40].

3 Conventional In Vitro Assays for Assessing Genotoxicity

Genotoxicity is the process by which agents interact with DNA and other cellular targets that control the integrity of the genetic material. This includes induction of DNA adducts, strand breaks, point mutations, and structural and numerical chromosomal changes. Tests for genotoxicity are considered short-term in nature and are an integral part of product safety assessment. The purpose of conducting these studies can be broadly divided into four categories: (1) to identify the potential of a particular environment (chemical/electromagnetic) to cause damage to a cell's DNA; (2) to screen compounds for potential carcinogenic activity during early drug development to help select and/or prioritize potential leads; (3) to obtain a complete genotoxicity database on promising new molecules to support global regulatory requirements; and (4) to generate complementary data for better understanding of existing results.

Genotoxicity tests can be defined as in vitro or in vivo tests designed to detect compounds that induce genetic damage directly or indirectly by various mechanisms. These tests should enable hazard identification with respect to damage to DNA and its fixation. Fixation of damage to DNA in the form of gene mutations, larger scale chromosomal damage, recombination, and numerical chromosome changes is generally considered as essential for heritable effects [41]. Compounds or environmental samples which are positive in tests that detect such kinds of damage have the potential to be rodent and/or human carcinogens and/or mutagens [2, 3, 10, 11, 41, 42].

3.1 Assays Based on DNA Damage Induction

Advances in genomics and proteomics over the past decade have had a great impact on the process of investigating biomolecules (DNA, RNA, and proteins) in living organisms. A series of mammalian test systems is currently being used to demonstrate an agent's ability to interact with cellular DNA and cellular response is measured after short-term incubation with moderate to high doses of agents. There is an increasing body of evidence that the ability to repair DNA damage is an important characteristic of both cells and organisms and can serve as a marker for genetic instability, cancer risk, and environmental and occupational hazard as well as in chemo- and radiotherapy [43–45]. Detection and measurement of DNA damage belongs, without any doubt, to the most important subjects in genetic toxicology, especially when induction and repair of damage has to be correlated with biological endpoints (Fig. 2), such as cell inactivation and cellular recovery [47–49]. Accordingly, sensitive methods are needed which result in dose dependent effects.

A variety of experimental techniques have been developed to detect DNA damage in cell populations. Most methods can only detect the average amount of DNA damage over thousands to millions of cells, while only few reveal the heterogeneity of DNA damage within a sample, reflecting the fact that some cells experience extensive damage whereas others display no damage at all.

Large-scale assays make use of DNA isolated from exposed population prior to or while extracting DNA damage information from these samples. However, there are many biological circumstances that require the use of small cell samples. Such techniques rely upon the evaluation of DNA damage at the level of single cells.

3.1.1 Measurements of Adducts or Modified Bases

Antibodies have been developed against a wide array of carcinogen-DNA adducts as well as UV-damaged or oxidized bases. Their sensitivity for the detection of DNA damage in humans has been demonstrated in numerous studies. The first antisera against modified DNA recognized the alkylated adducts of several bases including O^6-methyl- and O^6-ethylguanine, 7-methylguanine, and N^6-methyladenine (reviewed in [50]). For bulky chemical carcinogens, the first antibodies were directed against adducts of *N*-2-acetylaminofluorene [51] followed by antibodies recognizing adducts formed by BPDE [52]. This area has expanded rapidly over the past 10 years [53–55]. Many scientists have routinely distributed their monoclonal antibodies to other researchers, making the method of adduct detection readily

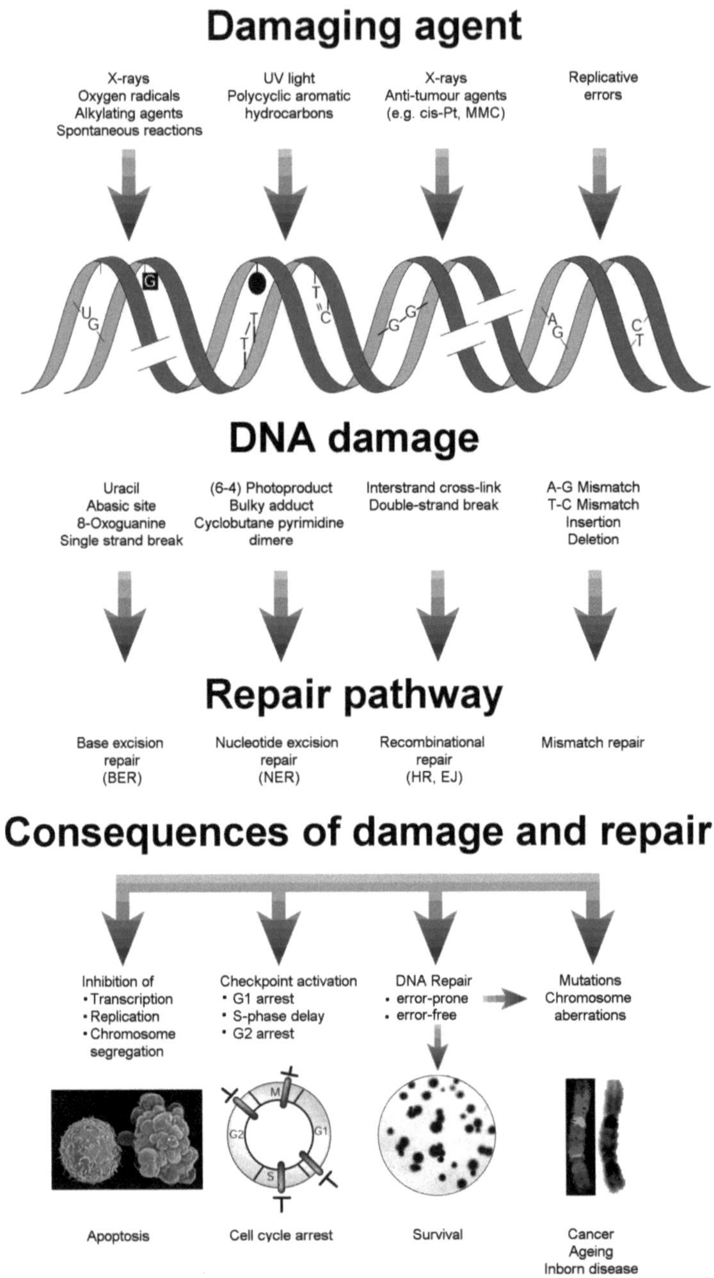

Fig. 2 DNA damage, repair mechanisms and consequences (modified from [46]). Common DNA damaging agents lead to a series of different DNA lesions. Acute effects of DNA damage on cell-cycle progression lead to checkpoint activation and transient arrest in the G1, S, G2, and M phases until DNA lesions are removed from cellular DNA by the action of repair mechanism. Long-term consequences of DNA injury include permanent changes in the DNA sequence in form

Table 1 Some commercially available antibodies against damaged DNA bases and adducts

DNA damage	Product name	Clone	Application	Supplier
(6-4) Photoproducts (6-4 PPs)	6-4PPs MAB	64M-2	ELISA/IC/IHC	CBI
5-Hydroxymethyluridine (5-HMU)	5-HMU PAB		ELISA	Abcam
5-Methyl cytidine (5-MC)	5-MC MAB	2E5 (3A3)	FC, SB, ICC, ELISA, IHC	Abcam
8-Hydroxy-2′-deoxyguanosine (8-OHdG)	8-OHdG MAB	7D7E4	ELISA, IHC	Abcam
Benzopyrene-7,8-diol-9,10-epoxide (BPDE)	BPDE-DNA MAB	5D11	IF and IHC	Santa Cruz
Cyclobutane pyrimidine dimers	CPDs MAB	TDM-2	ELISA/IC/IHC	CBI
Dewar photoproducts (Dewar PPs)	Dewar PP MAB	DEM-1	ELISA/IC	CBI
Malonaldehyde deoxyguanosine adduct (MADGA)	MADGA MAB		AP, WB	GenWay
Single stranded DNA	ssDNA MAB	F7-26	FC, ICC, IHC	Alx
Thioguanine (TG)	TG MAB	HYB 125-03	ELISA	Abcam
Thymine dimer (TD)	TD MAB	H3	ELISA, ICC, SB	Abcam

ELISA enzyme linked immune sorbent assay, *IC* immune chemistry, *IHC* immune histochemistry, *ICC* immune cytochemistry, *IF* immune fluorescence, *SB* southern blot, *FC* flow cytometry, *AP* affinity purification, *WB* western blot

CBI Cosmo Bio International Sales, Tokyo, Japan; Abcam, Cambridge, MA, USA; GenWay GenWay Biotech, San Diego, CA, USA; Santa Cruz, Santa Cruz Biotechnology, Santa Cruz, CA, USA; Alx, AXXORA, LLC, San Diego, CA, USA

accessible. Meanwhile, a series of antibodies against multiple damaged DNA bases and adducts are commercially available (Table 1). They can be used in a wide variety of methods of detection, such as radio immune assays (RIA), competitive enzyme linked immune assays (ELISA) using alkaline phosphatase or peroxidase-conjugated secondary antibodies for primary antibody detection and colored or fluorescent substrate, or immune histochemical detection (IHC) carried out on either fixed cells (e.g., lymphocytes or exfoliated oral or bladder cells) or tissue sections (frozen or paraffin) using fluorescent-coupled primary or secondary antibodies.

The use of immunological methods for monitoring DNA damage in human cells is well established, although specific antibodies must be developed to each adduct or class of adducts of interest. For screening of chemicals for their ability to produce DNA adducts which have not been characterized chemically, only the postlabelling method is applicable. In the analysis of DNA adducts by the ^{32}P-post-labelling assay [56], DNA containing adducts is isolated from treated cells and digested enzymatically to nucleotide 3′-monophosphates. Adducted nucleotides are then enriched prior to labeling. For enrichment of bulky and aromatic DNA adducts a butanol extraction and nuclease P1 treatment are the method of choice [57], while

Fig. 2 (continued) of mutations affecting single genes or chromosome aberrations which may involve multiple genes) and their biological effects. Abbreviations: *cis-Pt* and *MMC* cisplatin and mitomycin C, respectively, *HR* homologous recombination, *EJ* end joining

methyl- and ethyl-substituted products at the N7 position can be enriched as a group or class by an anion-exchange column chromatography [58]. Following the digestion and enrichment steps, adducted nucleotides are ^{32}P-labelled at the 5′-end, separated by thin-layer chromatography, detected by exposure to X-ray films or scanning by electronic autoradiography, and quantified by measuring the ^{32}P incorporation.

3.1.2 Measurements of Apurinic and Apyrimidinic Sites

Apurinic and apyrimidinic sites (AP sites) are one of the major types of DNA lesions generated by breakage in the glycosylic bond between the base and ribose. They are formed during the course of base excision and repair of oxidized, deaminated or alkylated bases. It has been estimated that about 2×10^5 base lesions are generated per cell per day. The level of AP sites in cells can be a good indicator of DNA lesion and repair against chemical damage and cell aging. AP sites can be detected using a biotin-containing Aldehyde Reactive Probe (ARP), which is a biotinylated reagent for the quantification of AP sites in damaged DNA. ARP reacts with aldehyde groups of AP sites formed when reactive oxygen species depurinate DNA, resulting in covalent linkage of biotin to these AP sites [59]. The biotin-tagged DNA can then be detected using common avidin-conjugated reporters such as avidin-coupled horse radish peroxidase. The ARP method is highly sensitive, enabling detection of about two AP sites per 1×10^7 nucleotides of DNA [60].

3.1.3 Measurements of DNA Breaks

Other types of DNA lesions are breaks occurring in the sugar phosphate backbone of DNA. DNA single-strand breaks (SSBs) are the commonest DNA breaks that arise directly from the attack of deoxyribose by reactive oxygen species in living cells [61]. Double-strand breaks (DSBs), in which both strands in the double helix are broken in close proximity, are particularly hazardous to the cell because they can lead to genome rearrangements [62]. Several methods for quantitative measurements of strand breaks have been described, which assess DNA damage either under neutral or alkaline conditions. In all cases of quasi-neutral conditions (pH 7.0–9.6), DNA remains in its double-stranded (ds) form, thus allowing only DSB detection [63].

Methods based on alkaline conditions make use of the fact that dsDNA unwinds in alkaline solutions, allowing the detection of DSBs, SSBs, and alkali-labile sites, damages which summarily are called DNA breaks. These free DNA ends, which are multiplied in cells that have been exposed to radiation or DNA-reactive agents, are the starting points from which unwinding begins [64, 65]. The velocity and thus the extent of denaturation depends on the experimental conditions. Under stringent alkaline conditions (pH > 12.6), DNA unwinding is completed within short time, resulting in pure single-stranded (ss) DNA. The reduction of molecular length reflects the amount of all induced DNA strand breaks. Under moderate alkaline conditions (pH 12.2–12.4) such alkaline treatment inactivates DNA-degradative enzymes, removes proteins from the DNA and, separates the two strands of the double-helix. DNA-strand interruptions, both metabolism- and damage-induced, are the starting points for unwinding under alkaline conditions. DNA unwinding as a time-dependent process must be stopped by neutralization after some time followed by immediate fragmentation of the DNA in order to prevent renaturation. As such, unwinding results in only partially unwound DNA for moderate unwinding times and the velocity of unwinding or the amount of unwound DNA reflects the amount of overall DNA strand breaks. If damaged DNA is only partially unwound, ssDNA and dsDNA fractions are separated by hydroxyapatite chromatography [66] or differentially stained using fluorescence dyes with preferential binding to dsDNA. This fluorometric analysis of DNA unwinding (FADU assay), makes use of the fluorescent dyes ethidium bromide, DAPI, bisbenzimide, or picogreen [67–72]. Complete DNA denaturation is required for analytical methods which determine molecular size. Velocity sedimentation requires lysis of cells directly on top of an alkaline sucrose gradient [73, 74] followed by low speed ultracentrifugation and determination of the sedimentation position of damaged DNA compared to that of undamaged DNA. The alkaline filter elution assay [75, 76] as well as the alkaline version of the pulsed-field gel electrophoresis [77] determine parameters related to the molecular size distribution of broken DNA. In filter elution, molecular size is reflected by the velocity of DNA passing through a narrow-sized pore filter, while electrophoresis measures the amount of DNA released from the gel plug (fraction of activity released: FAR) as well as the size distribution of the DNA entering the gel to compare the effects of exposure.

Sucrose density gradient centrifugation as well as filter elution and pulsed-field gel electrophoresis may be performed in the alkaline as well as in the neutral version in order to detect SSBs and DSBs, respectively [78].

However, none of these methods is able to give absolute number for strand breaks without calibration. Neither are any of these methods, with the exception of immune histochemical assays using antibodies, able to locate definitively the lesions in individual cells or in tissue made up from a variety of cell subpopulations which represent different target sensitivities.

Over the past few decades, single cell gel electrophoresis (SCGE) which is also known as comet assay has become one of the standard methods for assessing DNA damage [79–81]. It is technically simple, relatively fast, cheap, and DNA breaks can be investigated in virtually all mammalian cell types without requirement for cell culture. It is sometimes used without knowing how it works or what sort of information it provides, but the fact that it is so successful at demonstrating DNA damage is enough to justify its use.

In the comet assay's simplest form, cells are embedded in agarose on a microscope slide and immersed in a lysis solution to remove lipids and proteins. DNA within the cells is subjected to electrophoresis (0.8 V cm^{-1}, 15 min). Whereas DNA in control cells remains within the cell core, broken DNA moves from this core to the anodic site, forming an image of a comet. After electrophoresis, DNA migration is visualized in a fluorescence microscope using a fluorescent dye. Individual images can then be digitized and analyzed for informative properties, such as size and distribution of fluorescence within the comet. These features give an indication of the number of strand breaks present in the cell. The comet assay can be performed under neutral as well as alkaline conditions. The neutral comet assay recognizes DSBs [82], while under alkaline conditions DNA migration is caused by strand breaks, alkaline labile sites, and transient repair sites [79–81].

The comet assay is easy to perform and adaptable to the study of many questions. It has high sensitivity (0.1–1 breaks per 10^9 Da: 100–1000 breaks per mammalian cell) and works at low agent concentrations [83]. Nevertheless, is a laborious method, because many cells have to be monitored and the data have to be related to graded calibration curves [84].

A modified neutral SCGE assay has been used in several epidemiological studies, mostly for studying oxidative damage [85, 86]. To this end, DNA is treated with lesion-specific enzymes that introduce breaks at sites of damage such as endonuclease (EndoIII) or T4 endonuclease V (EndoV), which are specific for oxidized pyrimidines and formamido pyrimidine glycosylase (Fpg), recognizing ring-opened purines [87] as well as 8-oxoguanine glycosylase (hOGG1) specific for 8-oxoguanine [88]. The DNA breaks introduced by these agents are detected by the SCGE assay. The other methods for strand break analysis which are described above can also be combined with the action of lesion-specific enzymes [89].

3.2 Assays Based on Mutation Induction

Mutation induction is frequently determined for animal or human exposure conditions as alterations of the cellular genomes. Subsequent tests determine the number of chromosome aberrations, micronuclei or sister chromatid exchanges [90–92]. Structural aberrations may be of chromosome or chromatid type changes. With the majority of chemical mutagens, induced aberrations are of the chromatid type, but chromosome type aberrations also occur. An increase in polyploidy may indicate that a mutagen has the potential to induce numerical aberrations. Chromosome mutations and related events are the cause of many human genetic diseases. There is substantial evidence that chromosome mutations and related events cause alterations in oncogenes and tumor suppressor genes of somatic cells and are involved in cancer induction in humans and experimental animals. Numerical aberrations can also be determined in exposed cell cultures; however, the method requires trained personnel and is time and labor intensive.

Gene mutation assays in cultured mammalian cells may be used to detect gene alterations induced by chemical substances, and yet they are rapid and simple to carry out when compared to the use of whole mammals.

Methods to assess mutagenicity of physical and chemical agents depend on growing colonies in a selective medium. Many mammalian cell gene mutation assays are available, but only four cell lines and three genetic loci are well validated and widely used [93]. These are V79 and CHO Chinese hamster cells, human lymphoblastoid TK6 cells, and mouse lymphoma L5178Y cells, and the following genetic loci: hprt, tk, and the cell membrane Na+/K+ ATPase.

The gene encoding hypoxanthine guanine phosphoribosyltransferase (HPRT) has attracted considerable interest for mutagenesis studies because of its utility as a selectable marker. The enzyme, which is a member of a family of phosphoribosyl-transferases, constitutes a purine salvage or reutilization pathway that utilizes hypoxanthine and 5-phosphoribosyl-1-pyrophosphate (PRPP) to form AMP and GMP. The HPRT gene is located on the X chromosome, and cells are relatively tolerant to large genetic changes in this gene, as they can use a second de novo pathway for purine synthesis. In male cells there is a hemizygous situation and, accordingly, recessive mutations can be assessed. Whether or not the enzyme is expressed provides the basis for the sensitive selection system that permits selection of $HPRT^-$ cells using the toxin 6-thioguanine [94]. While non-mutated cells die in presence of the selective agent, mutated cells survive. In the HPRT assay V79 Chinese hamster cells are exposed to the test substance, both with and without metabolic activation, for 2–4 h or 24–72 h and subcultured to determine cytotoxicity and to allow phenotypic expression prior to mutant selection. Mutant frequency is determined by seeding known numbers of cells in a medium containing the selective agent 6-TG to detect mutant cells, and in a medium without a selective agent to determine the cloning efficiency (viability). After a suitable incubation time (ca. 10 days), colonies are stained and counted. The mutant frequency is derived from the number of mutant colonies in selective medium and the number of colonies in nonselective medium.

Human lymphoblastoid TK6 and mouse lymphoma $L5178Y^{TK+/-}$ cells detect gene mutations (point mutations) and chromosomal events (deletions, transloca-tions, mitotic recombination/gene conversion, and aneuploidy) at the thymidine kinase locus [95–97]. Mutant cells, which are deficient in the enzyme TK necessary for phosphorylation of thymidine to thymidine monophosphate, are resistant to the cytotoxic effect of pyrimidine analogues, such as trifluorothymidine. The mouse lymphoma assay (MLA) detects mutations known to be important in the etiology of cancer and other human genetically mediated illnesses.

The third marker used for mutagenicity studies is the gene coding for Na^+-K^+–ATPase. In nonmutated cells, the Na^+-K^+–ATPase mediates the active transport of Na^+ and K^+ across the plasma membrane, a function that can be inhibited specifically by the cardiac glycoside ouabain, which is toxic to mammalian cells in culture. Resistance to its cytotoxic effects have been obtained in rodent and human cell strains, which are thought to have a mutation affecting the Na^+-K^+–ATPase so that their enzyme has a lower affinity for ouabain than the enzyme of wild-type cells [98, 99].

The MLA based on the TK gene and the V79-HGPRT assay effectively measure specific types of mutations but are limited in sensitivity [100] by the requirement that flanking genes on the chromosome remain functional for cell survival. If the mutation extends beyond the mutated gene location, it may then cause cell death and the mutation is not scored. This is especially true in the HGPRT assay because the gene is located on the X-chromosome and flanking genes may not be rescued by a homologous chromosome. Large deletions, for example, are likely to kill the cell and alter the accurate mutant yield induced by a genotoxic agent, thereby reducing the assay sensitivity [101].

In light of these difficulties, a mammalian cell mutation assay was designed around a CHO cell line (CHO AL) that stably incorporated a single copy of human chromosome 11, which encodes several cell surface proteins including the glycosylphosphatidylinositol (GPI)-anchored cell surface protein CD59 [102, 103] located at 11p13.5. As a consequence of a mutation in the CD59 gene, its expression is reduced or lost. The newest versions of the human-hamster hybrid (AL) test measures the CD59-mutant cell yields by quantifying the fluorescence of cells labeled with phycoerythrin-conjugated mouse monoclonal anti-CD59 antibody using flow cytometry [104, 105].

3.3 Assays Based on Cellular DNA Damage Response

As mutations in the cellular genomes arise not only from physical or chemical attack at DNA level but also from the action of repair systems, a series of genotoxicity assays consider investigations on DNA repair as suitable for identifying the action of genotoxins on cells. In this respect, DNA repair refers to a collection of processes by which a cell identifies and corrects damage to the DNA molecules that encode its genome. Two principles can be followed in respect to DNA damage repair studies: demonstration of the reduction of DNA damage and visualization of molecules integrated in DNA repair functions.

The first principle relies on the fact that the number of induced damages will be reduced by successive repair actions with increasing time after exposure. That means that any of the methods described in Sect. 3.1 also document DNA repair when the assay is performed after a certain exposure time. From this point of view, interest is focused on the difference between initial and residual (unrepaired) damage (Fig. 3). Different mathematical approaches can be used to describe repair kinetics over time, either based on DNA repair saturation [106] or on DNA damage decay [107–109].

Saturation kinetics emanates from the hypothesis that DNA repair reaches a distinct maximum dependent on dose and cell type. The kinetics plot then results in a hyperbolic graph mathematically defined at any point. This model works even with very high levels of repair, but results only in the half-life for the whole repair process considered. Decay kinetics, on the other hand, assume repair of different kinds of DNA damage to proceed with a certain half-life. Kinetics curves may accordingly be dissected into repair subkinetics according to differently operating repair pathways. This calculation method predicts that a certain fraction of residual damage will always remain. Accordingly, the model does not work if repair is nearly complete and DNA damage levels are located in the range of nonexposed cells.

The second screening principle acts on the hypothesis that repair leaves its own marks with regard to special functional and structural features. While the need for

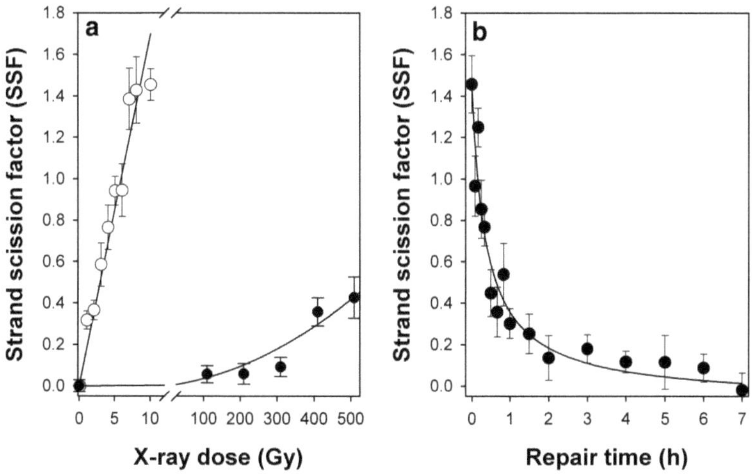

Fig. 3 DNA damage and repair measurements of Chinese hamster ovary K1 cells after irradiation with 200 kV X-rays. Damage detection was performed using the fluorescent analysis of DNA unwinding approach. Dose dependency (**a**) for measurements immediately after irradiation (*open circles*) and after 24 h (*filled circles*) and repair kinetics (**b**) for cells exposed to 10 Gy of X-rays

DNA precursor synthesis for DNA repair has long been known, the need for modification of a series special signaling molecules and repair factors was only discovered in the last few decades.

Unscheduled DNA synthesis (UDS) is based on the fact that, during certain repair steps, DNA precursors are integrated in damaged DNA. Cells treated with a genotoxin are supplemented with labeled DNA precursors, preferably with tritiated thymidine, which is incorporated into the damaged DNA in the course of DNA repair. In nonproliferating cells, the amount of the incorporated precursor is a direct measure for the repair capacity of the cells [110, 111]. In proliferating populations, S-phase cells and mitoses can be differentiated from cells performing repair by autoradiography. The method is sensitive and could be used to assess a broad variety of lesions, but its limitation is that it is applicable only for lesions which are removed by the nucleotide excision repair (NER) pathway, where the repair of the damage results in the de novo synthesis of uniformly long DNA stretches. The method has been used to assess the DNA repair activity of human malignant lymphoid cells.

The number of signaling molecules and repair factors is increasing from year to year, starting with the discovery of the function of p53 as guardian of the genome [112–114] and followed by the identification of the two master kinases, ataxia telangiectasia mutated protein (ATM) and ataxia telangiectasia and Rad3 related protein (ATR). ATM responds to DNA DSBs and disruptions in chromatin structure, whereas ATR primarily responds to stalled replication forks [115]. These kinases phosphorylate a series of downstream targets in signal transduction cascades (Table 2) which are involved in DNA damage repair, cell cycle control, and apoptosis.

Table 2 A selection of DNA signaling molecules and repair related factors against which antibodies are commercially available

Genes	Protein function
53PB1	Checkpoint signaling during mitosis. Enhancement of TP53-mediated transcriptional activation. Binding to sites of DNA damage in DNA damage domains [116]
ATM	Serine/threonine protein kinase which activates checkpoint signaling upon double strand breaks (DSBs), apoptosis and genotoxic stresses. Phosphorylates Ser-139 of histone variant H2AX/H2AFX at double strand breaks (DSBs), thereby regulating DNA damage response mechanism [115]. Also involved in signal transduction and cell cycle control
BRCA1	Cell cycle arrests in both the S-phase and the G2 phase of the cell cycle [117]. Transcriptional regulation of P21 in response to DNA damage. Required for FANCD2 targeting to sites of DNA damage. Disease: breast cancer susceptibility
CDKN1A P21CIP1 WAF1	Inhibitor of cellular proliferation in response to DNA damage. Binds to and inhibits cyclin-dependent kinase activity, preventing phosphorylation of critical cyclin-dependent kinase substrates and blocking cell cycle progression [118]. Phosphorylation of Thr-145 by Akt or of Ser-146 by PKC impairs binding to PCNA
Cyclin E	Control of the cell cycle at the G1/S transition. Interacts with a member of the CDK2/CDK protein kinases [119]. Interacts with Rb binding protein 3 and Rb-like protein 1. Phosphorylated upon DNA damage, probably by ATM or ATR
DNA-PKCs DNAPK1 PRKCD XRCC7	Serine/threonine-protein kinase that acts as a molecular sensor for DNA damage. Phosphorylated upon DNA damage, probably by ATM or ATR. Involved in DNA nonhomologous end joining (NHEJ) required for double-strand break (DSB) repair and V(D)J recombination. Phosphorylates Ser-139 of histone variant H2AX/H2AFX, thereby regulating DNA damage response mechanism [120]
GADD45A DDIT1	Binds to proliferating cell nuclear antigen. Might affect PCNA interaction with some cell division protein kinase complexes; stimulates DNA excision repair in vitro and inhibits entry of cells into S phase [121]
H2AX	Variant histone H2A which is phosphorylated on Ser-139 (to form γ-H2AX) in response to DSBs, generated by exogenous genotoxic agents and by stalled replication forks. Also occurs during meiotic recombination and immunoglobulin class switching. Interacts with many proteins required for DNA damage signaling and repair, such as MDC1, TP53BP1, BRCA1, and the MRN complex. Phosphorylation of Ser-139 is mediated by ATM and PRKDC. Dephosphorylation of Ser-140 by PP2A is required for DNA DSB repair [122]
KI-67 MKI67	Thought to be required for maintaining cell proliferation. Predominantly localized in the G1 phase in the perinucleolar region, in the later phases it is also detected throughout the nuclear interior, being predominantly localized in the nuclear matrix. In mitosis, it is present on all chromosomes [123]
Ku80/Ku86 XRCC5 Ku70 XRCC6	Single stranded DNA-dependent ATP-dependent helicase. Involved in DNA nonhomologous end joining (NHEJ) required for double-strand break repair and V(D)J recombination. The Ku p70/p86 dimer is probably involved in stabilizing broken DNA ends and bringing them together [124]. Phosphorylation on serine residues by PRKDC may enhance helicase activity

(*continued*)

Table 2 (continued)

Genes	Protein function
MDM2	Inhibition of TP53/p53 and TP73/p73-mediated cell cycle arrest and apoptosis. Permits the nuclear export of p53 and targets it for proteasome-mediated proteolysis [125]
MRE11	Plays a central role as a component of the MRN complex in DSB repair, DNA recombination, maintenance of telomere integrity and meiosis. Required for DNA damage signaling via activation of the ATM kinase [126]
	Component of the MRE11/RAD50/NBN (MRN complex). Involved in DSB repair, DNA recombination, maintenance of telomere integrity, cell cycle checkpoint control and meiosis [127]. NBN recruits PI3/PI4-kinase family member ATM, ATR, and probably DNA-PKcs to the DNA damage sites and activates their functions [126]
CDKN2A P16	Cycle arrest in G1 and G2 phases [128]. Binding to MDM2, thereby blocking MDM2- induced degradation of p53 and enhancing p53-dependent transactivation and apoptosis
PCNA	Auxiliary protein of DNA polymerase δ. Involved in the control of eukaryotic DNA replication during elongation of the leading strand [129]
Rad51C Rad51D	Involved in Homologous recombination repair (HRR) pathway of double-strand DNA breaks [130]
P53 TP53	Acts as a tumor suppressor in many tumor types; induces growth arrest or apoptosis depending on the physiological circumstances and cell type. Involved in cell cycle regulation as a trans-activator that acts to negatively regulate cell division by controlling a set of genes required for this process. Phosphorylated upon DNA damage, probably by ATM or ATR [112, 131, 132]

One of the key molecules in DNA damage signaling is the histone variant H2AX, which is involved in DNA repair and the maintenance of genomic stability. H2A.X is a histone variant that becomes phosphorylated on serine 139 by ATM or ATR early in the response to DNA double strand breaks. Its functions are attributed DSB repair, both in homologous recombination and nonhomologous end joining DNA repair pathways [122]. H2AX is thought to have a critical function in the recruitment of DNA repair factors and DNA damage-signaling proteins, while hyperphosphorylation of H2AX may be linked to chromatin fragmentation prior to apoptosis. Antibodies directed against the phosphorylated variant of H2AX can be used to visualize DNA damage loci (Fig. 4) by immune-fluorescent techniques [133]. Colocalization studies using antibodies directed against different DNA signaling molecules and repair related factors reveal insight into the complicated network of DNA damage responses [126, 134–136] leading to cell survival or cell death.

A considerable body of work on cellular DNA damage response has focused on the interplay between life and death mechanisms, namely apoptosis and antiapoptosis. Two pathways controlled by the ATM protein kinase, the master regulator of the cellular network induced by DNA double strand breaks, have been reported to be of special interest. It has been shown that pro- and antiapoptotic signals were

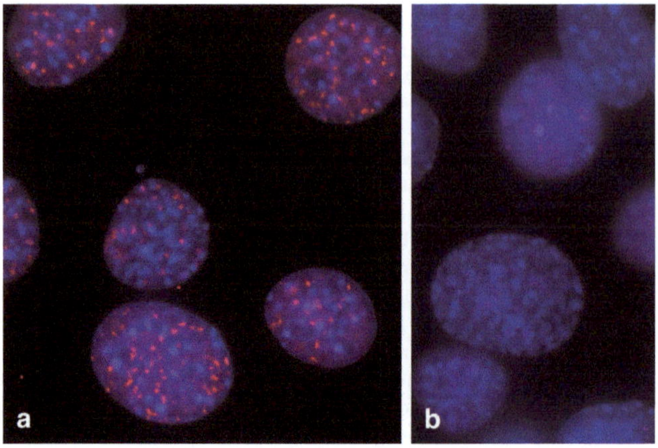

Fig. 4 DNA damage response as measured by γ-H2AX focus-formation in mouse osteoblastic cells (OCT-1) after exposure with 1 Gy of 160 kV X-rays (**a**) and unirradiated cells (**b**). Cells were fixed with paraformaldehyde 1 h after exposure and immuno-stained with antiphospho-histone H2AX (S139) antibody (AF2288, R&D Systems) and phycoerythrin-coupled goat anti-rabbit IgG (F0110, R&D Systems) for γ-H2AX focus visualization (*red*). DNA is counterstained with DAPI (*blue*)

simultaneously induced by DNA damaging agents, with the proapoptotic pathway mediated by p53 targets, and the prosurvival pathway by NF-κB targets [137].

The transcription factor p53 plays a central role as cell cycle regulator in maintaining the integrity of genome, thereby acting as a tumor suppressor [138].Within short time after DNA-damaging events, p53 protein accumulates in the cell nucleus after posttranscriptional phosphorylation [131]. It mediates cell cycle arrest and apoptosis by activating the transcription of specific genes involved in growth regulation and apoptosis [139]. One of the major targets of p53 involved in cell cycle control is the cyclin-dependent kinase inhibitor 1A ($p21^{CIP1/WAF1}$), encoded by the gene CDKN1A. Its upregulation is tightly controlled in response to a variety of stress stimuli and leads to a possible G_1 arrest [140] and enhanced DNA repair as a consequence; its downregulation may be correlated with increased apoptotic cell death.

The pathway that antagonizes apoptosis and hence favors cell survival is controlled by the transcription factor NF-κB [132]. Members of the NF-κB family share a Rel homology domain (RHD), which mediates DNA binding, dimerization, and interactions with specific inhibitory factors (the IκBs) which retain NF-κB dimers in the cytoplasm. Upon stimulation NF-κB is activated, mostly through IκB kinase dependent (IKK-dependent) phosphorylation and subsequent IκB degradation. The liberated NF-κB enters the nucleus, where it regulates transcription of a large group of genes, encoding cytokines, growth factors, cell adhesion molecules, and pro- and antiapoptotic proteins [141]. This control is crucial to a wide range of biological processes, including oncogenesis and cancer chemo-resistance. During inflammatory response, NF-κB activation protects cells from apoptosis by suppression of the Jun N-terminal kinase (JNK) cascade [142]. Recently, the ATM/NEMO complex was identified as the long searched for "nuclear to cytoplasm" signal leading to IKK activation and subsequent NF-κB liberation upon DSB induction [143].

Assessment of mechanisms leading to cell death and survival as well as measurement of cell death itself are certainly important for toxicity measurements in the pharmaceutical industry. A considerable body of work on apoptosis has focused on the mitochondria and it is now widely accepted that this organelle is fundamental to the biochemistry of apoptosis [144]. Various biochemical pathways involved in the apoptotic cascade converge on the mitochondria, resulting in changes to membrane potential and the release of cytochrome c into the cytosol. Cytochrome c then initiates the activation of caspases, which mediate intracellular digestion [145]. The central role of the mitochondria in apoptosis, combined with the widely reported findings that apoptosis is an energy-requiring process, lead to investigation of subtle changes in energy metabolism. From the magnitude of the ratio of cellular ADP: ATP content, cells undergoing apoptosis and necrosis in cell cultures can be discriminated [146].

4 From Small Scale Assays to High Content Screening in Toxicology

Currently used toxicity testing schemes based on mammalian cells have taken years for development and validation. Although in vitro mammalian cell culture systems have been implemented for testing strategies, most of them are still practiced on a small scale as is the usual case for basic cell biological research. Small-scale cell biology is defined as the use of research tools, such as imaging-microscopes, flow cytometers or plate readers, for collecting data from a small number of experimental samples, usually from a few hundreds to some thousands of cells per experiment. Microscopy-based and cell-based systems typically require substantial amounts of time to conduct at significant costs. Because of the manual nature, such assays have suffered from low-throughput, and thus their use has traditionally been restricted to late stages of toxicity screening. Using technologies available in 1995 would take the equivalent of 20 years to screen 1 million compounds for their toxicity.

In contrast to small-scale screening methods, high throughput screening (HTS) is characterized by its simplicity, rapidity, low cost, and high efficiency. Methods of HTS are basically analytical biochemistry methods (photometry, electrophoresis, kinetic assay, radioisotopes, immunoassay, descriptive statistics, regression analysis, etc.). Various technologies, especially the novel technologies such as fluorescence, nuclear-magnetic resonance, affinity chromatography, surface plasmon resonance, and DNA microarray, are now available. The screening of more than 100,000 samples per day is already possible. This phenomenal progress has been achieved through assay miniaturization, parallel processing, and innovations in hardware and assay technologies. The parameters for a successful HTS assay are now well established. Assays should facilitate miniaturization to 96, 384, or 1536 well formats without detrimental effects on robustness, reproducibility, and statistical significance of the results. To function within these parameters on laboratory robots, homogeneous "mix and measure" assays are required as they avoid filtration, separation, and washing steps that are time consuming and difficult to automate [147].

Additionally, a certain number of key requisites have to be fulfilled in order to achieve readiness for marketing of alternative testing procedures. Different points need to be addressed for the evaluation of an in vitro assay for HTS, in particular validation in terms of intra- and interlaboratory reproducibility, high predictive power to guarantee correct risk assessment decisions, relevance of the selected endpoint to the type of compounds that are intended to be tested, simplicity, and low cost and benefit ratio [148]. Furthermore, issues such as signal-to-background ratio, reproducibility, and speed are paramount when assays are applied to HTS.

Cell-based assays for HTS fall into three broad categories: second messenger assays that monitor signal transduction following activation of cell-surface receptors; reporter gene assays that monitor cellular responses at the transcription or translation level; and cell proliferation assays to monitor the overall growth or death response of cells to external stimuli.

As already pointed out, in vitro toxicity assays have relied on a number of end points uncovered by different reporting molecules or chemical reactions leading to a measurable signal. Colorimetric, as well as luminescence- or fluorescence-based techniques are likely to be among the most important detection approaches suitable for HTS. Testing schemes for virtually all major toxicological endpoints described in section 2 have the potential for miniaturization and automation to provide a quick and more in-depth interpretation of compound-induced toxicity.

HTS-based cytotoxicity assays are now well established and analyze the effects of compounds on a number of parameters, such as nuclear size, mitochondrial membrane potential, intracellular calcium levels, membrane permeability, and cell number, offering increased specificity and selectivity for toxic events and allowing a higher level of prediction for future in vivo testing.

Miniaturized colorimetric assay systems have been described for the neutral red uptake test [149] and the tetrazolium salt reduction tests using MTT [150] and the newer salts MTS [151] and XTT [152]. Over recent years the focus has thus tended towards either fluorescent or luminescent endpoints, which facilitate much more simple and convenient assay procedures [153]. Luminescent HTS suitable assays are on the market for determination of ATP [23, 153] as well as for measurements and discrimination of apoptosis [154] and necrosis [146] and for determination of protein degradation mediated through the ubiquitin-proteasome pathway [155]. Miniaturized assays for DNA damage detection based on fluorescence readouts from 96-well plates have been described for measurements of DNA strand break using the micro-scale variant of the FADU test [156], the fast micro method [157], and the SCGE [158, 159]. A fast DNA damage assay for the induction of DNA DSBs based on two-color fluorescent staining of histone variant H2AX phosphorylated at serine 139 and of DNA is available from ActiveMotive Inc. (Cat. # 18030).

The fluorescence can be read out either by flow cytometry or by a microplate fluorescence reader. However, despite recent developments with high-speed machines [160], flow cytometry is currently not suited to the demands of high-throughput cytotoxicity testing. Furthermore, as H2AX phosphorylation takes place at metaphase in nontreated and treated cells [161], microplate readouts, others than microscopic single cell analyses, do not discriminate between the number of dividing cells in metaphase and DNA damage. Based on the use of microplate fluorescence readers, each assay described in Table 1 can be accommodated for HTS.

The drive towards cell-based HTS requires large quantities of individual cell types or different cell lines derived from various tissue origins available on demand. The necessity to maintain large quantities of many live cell cultures at a given time represents a bottleneck for screening as unforeseen delays in assays' scheduling on laboratory robots may be time consuming and costly. This drawback is fuelling the drive towards high content screening (HCS) that facilitates multiple analyses on the same cell samples at the same time. HCS is a high-throughput technology that derives a maximum of information from single experiments. It involves labeling of individual cell constituents with different labels and applies sophisticated image processing algorithms to analyze cell images generated by automated fluorescence microscopy. The technique allows measurement of changes in object localization, intensity, texture or shape, thereby enabling the analysis of physiologically relevant cellular events such as cell or protein movements, shape changes, protein modification or gene expression. By imaging each cell individually, it becomes possible to consider less than 100 cells within a 96-well microplate [162–165]. This parallels the need for new functional screens to discover what happens within a cell when a molecule interacts with its target. Moreover, the recent advent of RNA interference (RNAi) technology allows the systematic analysis of phenotypic cellular effects in large-scale genetic screens. Although HCS will require a few more years to demonstrate its potential power in toxicity screening, it has already allowed the automation of the widely used micronucleus assay [166].

5 The Concepts of Reporter Systems for Toxicology

As the pharmaceutical and biotechnology landscapes are moving towards cell-based HCS assays, the development of new systems sets the focus on genetic engineering techniques for the identification and validation of possible targets molecules and pathways [167] as well as for the generation of the test systems [168]. An important outcome of the genome projects is the fact that basic cellular structures, pathways, and signaling principles relevant for toxicology and stress responses show a high degree of conservation among different tissue and species [169, 170]. Model organisms that are easily approachable by genetic, biochemical, and physiological means can thus play an important role in the design of screening-oriented systems. An emerging trend in this respect is the generation of genetically engineered cell lines that contain transfected elements that can be viewed as biological toxicity reporter systems in vivo and in vitro [148].

5.1 Requirements for Reporter Systems

Commonly used detection techniques utilize isotopic, colorimetric, fluorometric or luminescent enzyme substrates products. Genetic reporter systems are invaluable for studying regulation of gene expression, both by cis-acting factors (regulatory promoter and enhancer sequences) or trans-acting factors (transcription factors or exogenous regulators).

In most cases, the element under investigation (promoter, enhancer) is cloned together with the reporter gene in an expression vector, which is subsequently used to transfect cells. Three important components should be integrated in reporter plasmids: the reporter gene, the minimal promoter containing the TATA box to which general transcription factors bind, and the transcriptional regulatory sequences that confer inducibility on the promoter.

As regulatory sequences, upstream regions of environmental, chemical, or virus responsive genes are frequently used. For constitutive gene expression, viral (simian virus 40, SV40; cytomegalovirus, CMV) and mammalian (β-actin) promoters are convenient [171], because they include both the minimal promoter and multiple binding sites for enhancers. For inducible gene expression, promoters responsive to DNA damage induced by chemicals, heat, or radiation are suitable [172]. Such regulatory fragments are usually several hundred nucleotides in length, and include various elements other than the receptor binding site of interest, so that the possible effects of other signaling pathways should be taken into consideration. Instead of a long promoter region containing a complex array of various elements, chemically synthesized responsive elements can be inserted upstream of the reporter gene [173]. In this case, the minimal promoter sequence should be selected separately, and that from the herpes simplex virus thymidine kinase (HSV-TK) gene is frequently used for this purpose [174]. Inserted responsive elements are often used as multimers to potentiate the response, as more than two copies may be necessary to maximize response. The strategy based on element responsive on external stimulation is extremely advantageous, because a series of reporter plasmids corresponding to various receptors can be obtained by simply replacing the oligonucleotide cassette.

Quantification of the reporter protein yields indirectly provides information on the transcription activity of the receptor element under investigation. Quantification can take place by detecting the corresponding mRNA, the reporter protein, or by measuring the enzyme activity of the reporter protein. When the reporter system is selected, care must be taken to ensure that the reporter gene is not already endogenously expressed in the examined cells and that the gene does not influence the physiology of the transfected cells. Frequently used reporter genes [175, 176] include the *E. coli* enzyme chloramphenicol acetyl transferase (CAT), the lacZ for

E. coli β-galactosidase (β-GAL), and secreted human placental alkaline phosphatase (SEAP). The firefly luciferase gene has been used because of its high sensitivity compared with that of colorimetrical reporter enzymes. The recently described *Gaussia* luciferase (G*luc*) has several advantages over previous luciferases as it possesses a natural secretory signal of 16 amino acids that drives its secretion into cell medium [177], thus allowing luminescence measurements without cell lysis, and its codon-humanized version produces a 100-fold higher luminescent signal intensity compared to firefly luciferase [178]. Fluorescent proteins, such as enhanced green fluorescent protein (EGFP) and its multiple colored variants as well as coral reef fluorescent proteins [179, 180] are now the elements of choice (Table 3), especially for high-throughput screening.

Since transcriptional activation mediated by receptors requires participation of multiple protein factors, it is important to use cell lines expressing these factors adequately. The human breast cancer cell line MCF-7 is known to provide the factors necessary to respond to estrogen, and is a frequently used cell line for screening of (anti)estrogenic substances [182]. The human hepatoma cell line HepG2 is an excellent model to investigate mitochondrial (mt) toxicity because of the cell's high contents of organelles and mtDNA, and actually different investigators are indeed using such cells [183]. However, HepG2 cells show only negligible levels of drug-metabolizing and do not constitute a real alternative to primary hepatocytes for toxicity screening. To overcome these constraints, recombinant models heterologously expressing P450 enzymes (CYPs) in different host cells have been developed [184, 185]. The recently described hepatoma cell line HepaRG, which homologously expresses CYP1A2, 2B6, 2C9, 2E1, and 3A4 at

Table 3 Properties of some fluorescent protein variants (modified from [181])

Class	Protein	Excitation (nm)	Emission (nm)	Brightness	Photo-stability	Oligomerization
Far-red	mPlum	590	649	4.1	53	Monomer
Red	mCherry	587	610	16	96	Monomer
	JRed	584	610	8.8	13	Dimer
	mStrawberry	574	596	26	15	Monomer
	DsRed	556	586	3.5	16	Monomer
	tdTomato	554	581	95	98	Tandem dimer
Orange	mOrange	548	562	49	9.0	Monomer
	mKO	548	559	31	122	Monomer
Yellow-green	YPet	517	530	80	49	Weak dimer
	mCitrine	516	529	59	49	Monomer
	Venus	515	528	53	15	Weak dimer
	EYFP	514	527	51	60	Weak dimer
Green	EGFP	488	507	34	174	Weak dimer
	Emerald	487	509	39	0.69	Weak dimer
Cyan	CyPet	435	477	18	59	Weak dimer
	mCFPm	433	475	13	64	Monomer
	Cerulean	433	475	27	36	Weak dimer
UV-excitable green	T-Sapphire	399	511	26	25	Weak dimer

high levels, could represent a unique model system for analyzing genotoxic compounds [186].

5.2 Currently Used Toxicity Reporter Systems

Revolutionary advances in molecular biology have offered exciting possibilities for the development of new toxicity screening reporter assays. Molecular biology has offered the exciting new DNA microarray technology to monitor expression changes of multiple genes and signaling pathways in parallel [187]. DNA microarrays were already used to investigate the expression of biological target molecules associated with either human disease or with different exposure scenarios, which then opens new and exciting opportunities for identifying key molecules and signatures for toxicity responses [167, 188]. Future advances in DNA microarray technology will undoubtedly result in the availability of affordable DNA chips that can be used to profile dozens and even hundreds of compounds in a short period of time as high throughput assays.

So far, a series of genes leading to defined expression profiles for toxic exposure conditions have proved to be suitable candidates to act as sensors for the development of receptor-reporter assays for genotoxicity. Among them are stress related genes and genes coding for DNA repair relevant proteins as well as for survival- and apoptosis-related pathways.

For determination of cytotoxicity, reporter assays are suitable, which express the reporter protein controlled by a constitutively regulated promoter, such as promoters for housekeeping genes or strong viral genes. A cell viability assay based on the B13 cell line transfected with the luciferase expression vector pHCMV-Luc which carries the firefly luciferase gene under control of the CMV promoter [189]. The assay requires the use of external added substrates and thus has little utility as in situ viability assay for HTS. Similar green fluorescent protein based assays were developed for biotechnology applications, as GFP from *Aequoria victoria* exhibits a spontaneous fluorescence in diverse cell types, is stable over time, and is independent of cofactors or substrates. Furthermore, GFP fluorescence can be easily monitored and quantified by flow cytometry or microplate-based fluorometric assays [180].

The recombinant rat mammary adenocarcinoma cell line R3230AC-pEGFP-N1 looks more directly at GFP fluorescence retention as a membrane integrity indicator, a generally accepted functional indicator of cell viability [190]. The cell viability assay based on EGFP expression by the CHO-AA8-pCX-xGFP cell line controlled by a strong hybrid promoter composed of a CMV enhancer and a β-actin/β-globin gene promoter [191, 192]. It describes counter-trends of EGFP-expression in cell populations exposed to UV-radiation. For cells with impaired cellular membranes, as is the case in apoptotic cells, EGFP fluorescence is undetectable due to a leak-out of the soluble EGFP molecules. On the other hand, cells with DNA damage are found to enrich EGFP dose-dependently, presumably as they pass into a cell cycle block and continue to synthesize proteins (as well as EGFP). This was shown by using inhibitors of DNA synthesis, such as hydroxyurea. In exponentially growing populations, treated with a cytotoxin, EGFP expression measurements can reflect the number of cells able to escape from DNA damage dependent cell cycle arrests [192]. Using the recombinant HEK-pEGFP-N1 cells system, the reduction of EGFP-yields as indicator for cytotoxic effects with increasing dose. Results from measurements with a microplate reader were shown to be in good agreement with results obtained from standard techniques for toxicity measurements, such as the CFA assay and the MTT test [193].

For determination of genotoxicity, reporter assays are suitable, which express the reporter protein in response to external stimuli. For such approaches, promoters need either to be induced or repressed as consequences of exposure to agents that damage DNA. Genes activated during early transcriptional responses include those involved in DNA repair, replication, and growth control.

The tumor suppressor p53 plays an important role in early transcriptional responses, as it is induced upon DNA damage essentially by post-translational regulatory mechanisms, which lead to a substantial increase of p53 levels, followed by its transcriptional regulation of its target genes.

The increase in p53 protein accumulation and its subcellular translocation from the cytoplasm to the nucleus can be visualized by a reporter system consisting of the human HEK293/pAQC23 cell line stably harboring a plasmid with a specific translational TP53::EGFP gene fusion [194]. The HEK293 cell line was chosen as a host for this assay as it contains a functionally stabilized p53 due to the expression of the transgenic adenoviral E1A oncoproteins. With this human reporter cell system, levels of p53 can be determined by fluorescence microscopy and by FACS analysis following treatment with several classes of genotoxic and carcinogenic compounds in a concentration-dependent manner. The tested DNA damaging agents, such as topotecan, doxorubicin, cisplatin, 4-nitroquinoline-*N*-oxide, mitomycin C, methylmethanesulfonate, *N*-methyl-*N*′-nitro-*N*-nitrosoguanidin, 2-aminopurine, ethyl methanesulfonate, aflatoxin B1, vincristine, hydrogen peroxide, and gamma irradiation, caused a significant increase in intracellular p53-EGFP levels. Nongenotoxic carcinogens and stress conditions, such as incubation with saccharin, phenobarbital, dimethylsulfoxide, hypoxia, heat and cold shock, and adenovirus infection, did not result in p53-EGFP accumulation.

Evaluating the transcriptional regulatory role of p53 can be done using the stable murine bone marrow stromal cell line D2XRIIGFP24 which expresses EGFP under control of a p53 responsive element [195]. The p53(15X)-d2-EGFP reporter was constructed by inserting 15 copies of a p53 response element and a minimal TATA promoter (TATAA) from plasmid p53-Luc into the reporter plasmid pd2-EGFP-1. Recombinant D2XRIIGFP24 cells display low constitutive background fluorescence which is significantly enhanced in response to exposure to oxidative and nitrosative stress, agents that induced p53 protein levels.

Another target gene of p53, the p53R2 subunit of ribonucleotide reductase, is responsible for allocation of DNA nucleotides after DNA damage has occurred. Two reporter assays make use of its promoter, the p53R2-dependent luciferase reporter gene assay [196, 197] and the p53R2-dependent EGFP reporter gene assay [198]. MCF-7 or HepG2 cells transfected with the p53BS-Luc reporter plasmid containing three tandem repeat sequences for the p53 binding site derived from the intron 1 of human p53R2 gene were used to examine the genotoxic potential of more than 80 chemicals. Out of 43 Ames-positive chemicals, 40 induced luciferase activity; however, 8 Ames-negative chemicals also induced luciferase activity [197]. From experiments with 27 chemicals, whose genotoxic mechanisms are well known, it was found that the assay shows greater response to those genotoxic agents that induce DNA double strand break damage compared to those agents that cause other forms of DNA damage [196]. Human lung epithelial cells A549-pRRM2b-EGFP transfected with the pRRM2b-EGFP reporter plasmid containing a hybrid promoter from two tandem repeat sequences for the p53 binding site derived from the intron 1 of human p53R2 gene and the thymidine kinase minimal promoter revealed a good response after treatment with adriamycin, UVC-radiation, X-irradiation, and exposure to heavy ions of different relative biological weighting [198], all conditions which are known to induce DNA double strand breaks.

A reporter assay for transcriptional activity by the p53 opponent NF-κB has been described [199] which uses expression of the destabilized EGFP variant d2EGFP controlled by a synthetic promoter as a feature of NF-κB activation. The assay is based on HEK cells stably transfected with a vector carrying EGFP or d2EGFP under control of a synthetic promoter containing four copies of the NF-κB response element. It identifies conditions that are capable of driving cells to follow the NF-κ B triggered antiapoptotic pathway. The test has been validated to reflect the time course of NF-κB activation by the classical [200] as well as the DNA damage dependent pathway [201, 202]. The detection of NF-κB activation with d2EGFP as reporter molecule is suitable to follow the up- and downregulation of cellular stress response due to its short half-life of about 5–6 h [193]. However, this assay requires FACS analysis or fluorescence microscopy for measuring. Although the use of EGFP disables visualization of downregulated NF-κB activation, as a tribute to EGFP's long half-life of >90 h and its accordant accumulation this assay variant enables measurements in the microplate reader, thereby rendering it useful for high-throughput screening.

Other cellular assays based on the expression of reporter molecules have been developed for analyzing DNA damage response as a surrogate for genotoxicity. Among them is the DNA-damage inducible growth arrest inducing gene GADD45α, which was found to be the only GADD gene dependent on p53 or NF-κB activation. Several assays make use of its promoter. The CAT-Tox (L) assay [203] uses a GADD45α CAT fusion reporter stably integrated into the HepG2 cell line. The GreenScreen HC assay makes use of the p53-proficient human lymphoblastoid cell line TK6 bearing the upstream promoter region and regulatory genes sequences of the human GADD45α gene operatively linked to a human codon optimized GFP gene [204, 205]. Studying 75 well-characterized genotoxic and nongenotoxic compounds with diverse mechanisms of DNA-damage induction revealed that the assay responds positively to all classes of genotoxic damage with both high specificity and high sensitivity.

Another set of reporter assays reflecting transcriptional response upon exposure to DNA damaging agents is mediated by use of the recombinant human colorectal cancer cell line RKO in the CAT-Tox (D) assay [206]. Seven different reporter constructs derived from promoters and/or response elements were fused to the bacterial reporter gene choramphenicol acetyl transferase (CAT). The promoters used as response sensing elements are those from the genes coding for polymerase β, gadd45α, c-fos, or response elements for the action of p53, p153, TPA, and tissue-type plasminogen activator protein 1. The transcriptional responses are to be measured as a function of the accumulation of CAT protein using antibodies against CAT protein in a standard ELISA. Other assays make use of promoters for stress response genes, such as the murine 70 kDa heat shock protein Hsp70A.1 promoter [207], the human HSP70 promoter [208], or reporter plasmids driven by the cyclin A promoter, which is downregulated in response to toxic stress by decreasing phosphorylation of CREB and ATF-1 [209].

Several authors have taken advantage of EGFP variants to design apoptosis detection assays based on EGFP. These assays involve correct tracking of the nuclear pore membrane protein POM121 by a chimeric EGFP protein in living cells in contrast to apoptotic cells [210], or cleavage by activated caspases of a peptidic link between green and blue fluorescent proteins and evaluation of fluorescence resonance energy transfer [211] or inhibition of fluorescent reporter gene transcription [212]. Decreased EGFP fluorescence appeared to be an early and very sensitive marker of cell death; this might be due to the potential that redox changes occurring during the apoptotic process can be responsible for insufficient post-translationally regulated chromophore formation.

Several molecular approaches for observations of cellular DNA repair as indication for genotoxicity uses host cell reactivation of exogenous damaged plasmids [213]. For assessment of mismatch repair, which is responsible for detection and elimination of nonmatching nucleotides, a plasmid containing a corrupted EGFP gene by a T:G mismatch in its start codon can be used [214, 215]. This assay relies on restoration of the EGFP gene by mismatch repair mechanism and increased fluorescence yields in a chronic myelogenous leukemia (CML) cells.

A similar test was developed for measuring DSB repair using the plasmid pEGFP after damaging its marker gene for EGFP by UV-irradiation or other DNA damaging agents. As a reference, the plasmid pEYFP carrying the gene for the red-shifted variant EYFP is to be used. After cotransfection of both plasmids into HEK cells, the fluorescence of EGFP is determined relative to the fluorescence of EYFP using a fluorescence microplate reader to compensate for any differences of experimental variables [216]. A modified rapid dual fluorescent assay uses only one plasmid, the vector pREVY which contains two EGFP variant genes: a CMV promoter driven EGFP gene and a second CMV promoter driven EYFP [217]. The vector pREVY can be linearized at the unique SgfI restriction site immediately 5′ of the CMV promoter driving EGFP expression. After transient transfection of the linearized plasmid the ratio of EGFP and EYFP can be measured. From linear DNA, cells will only express EYFP but not EGFP. Unless correct cellular rejoining of the DSB has taken place, the ratio of EGFP/EYFP will be well below 1. The assay has the potential to be further developed to study the relationship between DSB repair fidelity and chemo-sensitivity.

Homologous recombination repair plays an important role in DSB repair and impairment of this repair mechanism may lead to loss of genomic integrity, which is one of the hallmarks of cancer. Agents which induce strand breaks that are to be removed by recombination events can be identified by the novel assay based on recombination between two fluorescent protein sequences in transfected plasmid DNA [218]. The plasmid construct contains an intact, emission-shifted, "blue" variant of EGFP (EBFP), with a 300 nucleotide stretch of perfect homology to a nonfunctional copy of the red-shifted green variant. In the absence of homologous recombination, only EBFP is present, but homologous recombination can create a functional GFP. The homologous regions in the plasmid were constructed in both the direct and the inverted orientation of transcription to detect possible differences in the recombination mechanisms involved. With this assay, genetic differences between host cell lines and different agents' signatures can be observed.

Currently, a series of signaling assay kits, such as the Cignal Reporter Assays, are on the market which provide rapid, sensitive, and quantitative assessments of pathway activation by measuring the activities of downstream transcription factors. The reporter systems available rely either on the dual-luciferase format or on the EGFP format. Although they contain preformulated transfection-ready reporter construct as well as positive and negative controls, the user has to decide on the suitable host cell line for transient transfections. This make such kits suitable tools for small scale use but not for HTS.

6 Future Perspectives and the Road to Acceptance

The strategy of reporter assays breaks new grounds for screening a large number of compounds rapidly for their potential toxic effects. And with the rapidly developing technologies it will soon be possible to advance further major improvements in the fields of toxicity screening. For example, genes encoding drug metabolizing enzymes, such as the human cytochrome P450 gene, can be directly introduced in cell lines or transgenic animals. Alternatively, specific genes can be deleted from the mouse genome and the consequences on toxicological response determined. Such applications have enormous potential for speeding up the drug discovery process and will undoubtedly become part of this process in the future.

In order to make the transition of new assays from small scale use in research laboratories to high throughput and HCS in pharmaceutical and environmental applications, the methods for the bioassays must be clearly defined and must meet widely accepted performance criteria. Regulatory adoption of new test methods is often a complex process, requiring test method validation, regulatory acceptance, and implementation. Testing procedures should first meet the criteria of national organizations and then meet international standards. Scientific validation is a necessary prerequisite for the regulatory acceptance of new toxicological tests, and for acceptance as Organization for Economic Co-operation and Development Test Guidelines. The validation process is designed to determine the operational characteristics of a test, that is, its reliability and relevance, in addition to its strengths and limitations. The reliability of a test is measured by its reproducibility. Its relevance is judged by its mechanistic relationship to the health effects of concern, and its ability to predict or identify those effects.

National and international organizations are currently developing new policies and standards for scientific practice to assure quality in implementation of in vitro methods. Test guidelines incorporating performance standards are being written to allow acceptance of proprietary test methods by regulatory agencies and to provide assurance that any in vitro system performs over time in a manner that is consistent with the test system as it was originally validated.

References

1. Foth H, Hayes A (2008) Concept of REACH and impact on evaluation of chemicals. Hum Exp Toxicol 27:5–21
2. Kirkland D, Aardema M, Henderson L, Müller L (2005) Evaluation of the ability of a battery of three in vitro genotoxicity tests to discriminate rodent carcinogens and non-carcinogens. I. Sensitivity, specificity and relative predictivity. Mutat Res 584:1–256
3. Kirkland D, Aardema M, Müller L, Makoto H (2006) Evaluation of the ability of a battery of three in vitro genotoxicity tests to discriminate rodent carcinogens and non-carcinogens. II. Further analysis of mammalian cell results, relative predictivity and tumour profiles. Mutat Res 19:29–42
4. Trevan JW (1927) The error of determination of toxicity. Proc R Soc Ser B Biol Sci 101:483–514
5. OECD (Organisation for Economic Cooperation and Development) (1987). Guideline for Testing of Chemicals – Acute Oral Toxicity. OECD Guideline 401
6. Cutler D (2001) Death of LD50. Trends Pharmacol Sci 22:62
7. OECD (Organisation for Economic Cooperation and Development) (2001). Updated guideline for testing of chemicals – acute oral toxicity-fixed dose method. OECD Guideline 420
8. OECD (Organisation for Economic Cooperation and Development) (2001). Updated guideline for the testing of chemicals – acute oral toxicity-acute toxic class method. OECD Guideline 423
9. OECD (Organisation for Economic Cooperation and Development) (1998). Guideline for the testing of chemicals – acute oral toxicity: up-and-down procedure. OECD Guideline 425
10. Goodman G, Wilson R (1999) Predicting the carcinogenicity of chemicals in humans from rodent bioassay data. Environ Health Perspect 94:195–218
11. Monro A (1996) Are lifespan rodent carcinogenicity studies defensible for pharmaceutical agents? Exp Toxicol Pathol 48:155–166
12. Rack L, Spira H (1989) Animal rights and modern toxicology. Toxicol Ind Health 5:133–143
13. Spielmann H (2002) A chair on alternatives? ALTEX 19:69–73
14. Balls M, Straughan DW (1996) The three Rs of Russell & Burch and the testing of biological products. Dev Biol Stand 86:11–18
15. Kreeftenberg JG (1999) Developments in the reduction refinement and replacement of animal tests in the quality control of immunobiologicals. Dev Biol Stand 101:17

16. van der Kamp MD (1996) Replacement, reduction or refinement of animal use in the quality control of veterinary vaccines: development, validation and implementation. Dev Biol Stand 86:73–76
17. Nath J, Krishna G (1998) Safety screening of drugs in cancer therapy. Acta Haematol 99:138–147
18. Puck TT, Monro A (1955) A rapid method for viable cell titration and clone production with HeLa cells in tissue culture: the use of X-irradiated cells to supply conditioning factors. Proc Natl Acad Sci USA 103:432–437
19. Puck TT, Monro A (1956) Action of X-rays on mammalian cells. J Exp Med 103:653–666
20. Johnson TR, Massey RJ, Deinhardt F (1972) Lymphocyte and antibody cytotoxicity to tumor cells measured by a micro-51 chromium release assay. Immunol Commun 1:247–261
21. Zarling JM, McKeough M, Bach FH (1976) A sensitive micromethod for generating and assaying allogeneically induced cytotoxic human lymphocytes. Transplantation 21:468–476
22. Mak S, Till JE (1963) Use of I-125-labeled 5-iodo-2′-deoxyuridine for the measurement of DNA synthesis in mammalian cells in vitro. Can J Biochem Physiol 41:2343–2351
23. Kangas L, Grönroos M, Nieminen AL (1984) Bioluminescence of cellular ATP: a new method for evaluating cytotoxic agents in vitro. Med Biol 62:338–343
24. Kuzmits R, Aiginger P, Müller MM, Steurer G, Linkesch W (1986) Assessment of the sensitivity of leukaemic cells to cytotoxic drugs by bioluminescence measurement of ATP in cultured cells. Clin Sci 71:81–88
25. Lundin A, Hasenson M, Persson J, Pousette A (1986) Estimation of biomass in growing cell lines by adenosine triphosphate assay. Methods Enzymol 133:27–42
26. Thore A, Anséhn S, Lundin A, Bergman S (1975) Detection of bacteriuria by luciferase assay of adenosine triphosphate. J Clin Microbiol 1:1–8
27. Untch M, Sevin BU, Perras JP, Angioli R, Untch A, Hightower RD, Koechli O, Averette HE (1994) Evaluation of paclitaxel (taxol), cisplatin, and the combination paclitaxel-cisplatin in ovarian cancer in vitro with the ATP cell viability assay. Gynecol Oncol 53:44–49
28. Kurbacher CM, Cree IA (2005) Chemosensitivity testing using microplate adenosine triphosphate-based luminescence measurements. Methods Mol Med 110:101–120
29. Mosmann T (1983) Rapid colorimetric assay for cellular growth and survival: application to proliferation and cytotoxicity assays. J Immunol Methods 65:55–63
30. Hayon T, Dvilansky A, Shpilberg O, Nathan I (2003) Appraisal of the MTT-based assay as a useful tool for predicting drug chemosensitivity in leukemia. Leuk Lymphoma 44:1957–1962
31. Shimoyama Y, Kubota T, Watanabe M, Ishibiki K, Abe O (1989) Predictability of in vivo chemosensitivity by in vitro MTT assay with reference to the clonogenic assay. J Surg Oncol 41:12–18
32. Borenfreund E, Babich H, Martin-Alguacil N (1990) Rapid chemosensitivity assay with human normal and tumor cells in vitro. In Vitro Cell Dev Biol 26:1030–1034
33. Borenfreund E, Shopsis C (1985) Toxicity monitored with a correlated set of cell-culture assays. Xenobiotica 15:705–711
34. Korzeniewski C, Callewaert DM (1983) An enzyme-release assay for natural cytotoxicity. J Immunol Methods 64:313–320
35. Skreb Y, Habazin-Novak V (1975) Reversible inhibition of DNA, RNA and protein synthesis in human cells by lead chloride. Toxicology 5:167–174
36. Rink H, Baumstark-Khan C, Partke HJ (1986) Radiation induced formation of giant cells (*Saccharomyces uvarum*): IV. Macromolecular synthesis and protein patterns. Radiat Environ Biophys 25:81–91
37. Baumstark-Khan C, Hellweg CE, Scherer K, Horneck G (1999) Mammalian cells as biomonitors of UV-exposure. Anal Chim Acta 387:281–287
38. Lockshin RA, Osborne B, Zakeri Z (2000) Cell death in the third millennium. Cell Death Differ 7:2–7
39. Hetz C (2008) Apoptosis, necrosis and autophagy: from mechanisms to biomedical applications. Curr Mol Med 8:76–77

40. Vermes I, Haanen C, Reutelingsperger C (2000) Flow cytometry of apoptotic cell death. J Immunol Methods 243:167–190
41. Dearfield KL (1995) Information requirements and regulatory approaches for heritable genetic risk assessment and risk communication. Mutat Res 330:35–40
42. Battershill JM, Fielder RJ (1998) Mouse-specific carcinogens: an assessment of hazard and significance for validation of short-term carcinogenicity bioassays in transgenic mice. Hum Exp Toxicol 17:193–205
43. Bulman RA, Dragani TA, Bouffler SD, Cox R (2006) Locations of mouse DNA damage response and repair loci, and cancer risk modifiers. DNA Repair 5:274–277
44. Naccarati A, Pardini B, Hemminki K, Vodicka P (2007) Sporadic colorectal cancer and individual susceptibility: a review of the association studies investigating the role of DNA repair genetic polymorphisms. Mutat Res 635:118–145
45. Pastwa E, Somiari RI, Malinowski M, Somiari SB, Winters TA (2009) In vitro non-homologous DNA end joining assays the 20th anniversary. Int J Biochem Cell Biol 41:1254–1260
46. Hoeijmakers JHJ (2001) Genome maintenance mechanisms for preventing cancer. Nature 411:366–374
47. Baumstark-Khan C (1993) X-ray-induced DNA double-strand breaks as lethal lesions in diploid human fibroblasts compared to chinese hamster ovary cells. Int J Radiat Biol 63:305–311
48. Baumstark-Khan C (1992) Effect of aphidicolin on DNA synthesis, PLD recovery and DNA repair of a human diploid cell line. Int J Radiat Biol 61:191–197
49. Blöcher D, Pohlit W (1982) DNA double strand breaks in Ehrlich ascites tumour cells at low doses of X-rays. II. Can cell death be attributed to double strand breaks? Int J Radiat Biol 42:329–338
50. Shuker DE, Farmer PB (1992) Relevance of urinary DNA adducts as markers of carcinogen exposure. Chem Res Toxicol 5:450–460
51. Poirier MC, Yuspa SH, Weinstein IB, Blobstein S (1977) Detection of carcinogen-DNA adducts by radiommunoassay. Nature 270:186–188
52. Poirier MC, Santella RM, Weinstein IB, Grunberger D, Yuspa SH (1980) Quantitation of benzo[a]pyrene-deoxyguanosine adducts by radioimmunoassay. Cancer Res 40:412–416
53. Baan RA, Zaalberg OB, Fichtinger-Schepman AM, Muysken-Schoen MA, Lansbergen MJ, Lohman PH (1985) Use of monoclonal and polyclonal antibodies against DNA adducts for the detection of DNA lesions in isolated DNA and in single cells. Environ Health Perspect 62:88
54. Santella RM (1988) Monitoring human exposure to carcinogens by DNA adduct measurement. Cell Biol Toxicol 4:511–516
55. Santella RM (1999) Immunological methods for detection of carcinogen-DNA damage in humans. Cancer Epidemiol Biomarkers Prev 8:733–739
56. Reddy MV (2000) Methods for testing compounds for DNA adduct formation. Regul Toxicol Pharmacol 32:256–263
57. Reddy MV, Randerath K (1987) 32P-postlabeling assay for carcinogen-DNA adducts: nuclease P1-mediated enhancement of its sensitivity and applications. Environ Health Perspect 76:41–47
58. Kumar R, Singh SK, Kole PL, Elmarakby S, Sikka HC (1995) Stereoselective metabolism of dibenz[a, h]acridine to bay-region diol epoxides by rat liver microsomes. Carcinogenesis 16:525–5130
59. Kurisu S, Miya T, Terato H, Masaoka A, Ohyama Y, Kubo K, Ide H (2001) Quantitation of DNA damage by an aldehyde reactive probe (ARP). Nucleic Acids Symp Ser 1:45–46
60. Nakamura J, Walker VE, Upton PB, Chiang SY, Kow YW, Swenberg JA (1998) Highly sensitive apurinic/apyrimidinic site assay can detect spontaneous and chemically induced depurination under physiological conditions. Cancer Res 15:222–225

61. Thompson LH, West MG (2000) XRCC1 keeps DNA from getting stranded. Mutat Res 459:1–18
62. Morgan WF, Corcoran J, Hartmann A, Kaplan MI, Limoli CL, Ponnaiya B (1998) DNA double-strand breaks, chromosomal rearrangements, and genomic instability. Mutat Res 404:125–128
63. Iliakis G, Blöcher D, Metzger L, Pantelias G (1991) Comparison of DNA double-strand break rejoining as measured by pulsed field gel electrophoresis, neutral sucrose gradient centrifugation and non-unwinding filter elution in irradiated plateau-phase CHO cells. Int J Radiat Biol 59:927–939
64. Daniel FB, Haas DL, Pyle SM (1985) Quantitation of chemically induced DNA strand breaks in human cells via an alkaline unwinding assay. Anal Biochem 144:390–402
65. Dusinská M, Slamenová D (1992) Application of alkaline unwinding assay for detection of mutagen-induced DNA strand breaks. Cell Biol Toxicol 8:207–216
66. Ahnström G, Erixon K (1973) Radiation-induced strand breakage in DNA from mammalian cells: strand separation in alkaline solution. Int J Radiat Biol 23:285–289
67. Baumstark-Khan C, Griesenbach U, Rink H (1992) Comparison of DNA strand break induction in CHO cells measured by alkaline elution and by fluorometric analysis of DNA unwinding (FADU). Free Radic Res Commun 16:381–389
68. Baumstark-Khan C, Hentschel U, Nikandrova Y, Krug J, Horneck G (2000) Fluorometric analysis of DNA unwinding (FADU) as a method for detecting repair-induced DNA strand breaks in UV-irradiated mammalian cells. Photochem Photobiol 72:477–484
69. Baumstark-Khan C, Nikandrova Y, Krug J, Hentschel U, Horneck G (1999) Adaptation of the fluorometric analysis of DNA unwinding (FADU) for the detection of repair-induced DNA strand breaks after UV-irradiation of mammalian cells. Trends Photochem Photobiol 6:141–152
70. Birnboim HC, Jevcak JJ (1981) Fluorometric method for rapid detection of DNA strand breaks in human white blood cells produced by low doses of radiation. Cancer Res 41:1889–1892
71. Elmendorff-Dreikorn K, Chauvin C, Slor H, Kutzner J, Batel R, Müller WE, Schröder HC (1999) Assessment of DNA damage and repair in human peripheral blood mononuclear cells using a novel DNA unwinding technique. Cell Mol Biol 45:211–218
72. Kanter PM, Schwartz HS (1982) A fluorescence enhancement assay for cellular DNA damage. Mol Pharmacol 22:145–151
73. Korba BE, Hays JB, Boehmer S (1981) Sedimentation velocity of DNA in isokinetic sucrose gradients: calibration against molecular weight using fragments of defined length. Nucleic Acids Res 9:4403–4412
74. McGrath RA, Williams RW (1966) Reconstruction in vivo of irradiated Escherichia coli deoxyribonucleic acid; the rejoining of broken pieces. Nature 212:534–535
75. Crump PW, Fielden EM, Jenner TJ, O'Neill P (1990) A comparison of the techniques of alkaline filter elution and alkaline sucrose sedimentation used to assess DNA damage induced by 2-nitroimidazoles. Biochem Pharmacol 40:621–627
76. Kohn KW, Grimek-Ewig RA (1973) Alkaline elution analysis, a new approach to the study of DNA single-strand interruptions in cells. Cancer Res 33:1849–1853
77. Sutherland JC, Monteleone DC, Mugavero JH, Trunk J (1987) Unidirectional pulsed-field electrophoresis of single- and double-stranded DNA in agarose gels: analytical expressions relating mobility and molecular length and their application in the measurement of strand breaks. Anal Biochem 162:511–520
78. Lu J, Kaeck MR, Jiang C, Garcia G, Thompson HJ (1996) A filter elution assay for the simultaneous detection of DNA double and single strand breaks. Anal Biochem 235: 227–233
79. Olive PL (2009) Impact of the comet assay in radiobiology. Mutat Res 681:13–23
80. Olive PL (2002) The comet assay. An overview of techniques. Methods Mol Biol 203: 179–194

81. Singh NP, McCoy MT, Tice RR, Schneider EL (1988) A simple technique for quantitation of low levels of DNA damage in individual cells. Exp Cell Res 175:184–191
82. Olive PL, Banáth JP (2009) Radiation-induced DNA double-strand breaks produced in histone-depleted tumor cell nuclei measured using the neutral comet assay. Radiat Res 142:144–152
83. Gedik CM, Ewen SW, Collins AR (1992) Single-cell gel electrophoresis applied to the analysis of UV-C damage and its repair in human cells. Int J Radiat Biol 62:313–320
84. De Boeck M, Touil N, De Visscher G, Vande PA, Kirsch-Volders M (2000) Validation and implementation of an internal standard in comet assay analysis. Mutat Res 469:181–197
85. Collins AR, Dobson VL, Dusinská M, Kennedy G, Stetina R (1997) The comet assay: what can it really tell us? Mutat Res 375:183–193
86. Collins AR, Dusinská M, Horská A (2001) Detection of alkylation damage in human lymphocyte DNA with the comet assay. Acta Biochim Pol 48:611–614
87. Piersen CE, McCullough AK, Lloyd RS (2000) AP lyases and dRPases: commonality of mechanism. Mutat Res 459:43–53
88. Girard PM, D'Ham C, Cadet J, Boiteux S (1998) Opposite base-dependent excision of 7, 8-dihydro-8-oxoadenine by the Ogg1 protein of *Saccharomyces cerevisiae*. Carcinogenesis 19:1299–1305
89. Mohsin Ali M, Kurisu S, Yoshioka Y, Terato H, Ohyama Y, Kubo K, Ide H (2004) Detection of endonuclease III- and 8-oxoguanine glycosylase-sensitive base modifications in gamma-irradiated DNA and cells by the aldehyde reactive probe (ARP) assay. J Radiat Res 45:229–237
90. Bonassi S, Ugolini D, Kirsch-Volders M, Strömberg U, Vermeulen R, Tucker JD (2005) Human population studies with cytogeneticbiomarkers: review of the literature and future prospectives. Environ Mol Mutagen 45:258–270
91. Norppa H, Bonassi S, Hansteen IL, Hagmar L, Strömberg U, Rössner P, Boffetta P, Lindholm C, Gundy S, Lazutka J, Cebulska-Wasilewska A, Fabiánová E, Srám RJ, Knudsen LE, Barale R, Fucic A (2006) Chromosomal aberrations and SCEs as biomarkers of cancer risk. Mutat Res 600:37–45
92. van Delft JH, Baan RA, Roza L (1998) Biological effect markers for exposure to carcinogenic compound and their relevance for risk assessment. Crit Rev Toxicol 28:477–510
93. Clements J (1995) Gene mutation assays in mammalian cells. Methods Mol Biol 43:277–286
94. O'Neill JP, Brimer PA, Machanoff R, Hirsch GP, Hsie AW (2009) A quantitative assay of mutation induction at the hypoxanthine-guanine phosphoribosyl transferase locus in Chinese hamster ovary cells (CHO/HGPRT system): development and definition of the system. Mutat Res 45:91–101
95. Kirkland DJ, Clements J (1998) Recommendations for spacing of test chemical concentrations in the mouse lymphomatk mutation assay (MLA). Mutat Res 415:159–163
96. Mitchell AD, Auletta AE, Clive D, Kirby PE, Moore MM, Myhr BC (1997) The L5178Y/tk+/− mouse lymphoma specific gene and chromosomal mutation assay aphase III report of the U.S. Environmental Protection Agency Gene-Tox Program. Mutat Res 394:177–303
97. Wang J, Heflich RH, Moore MM (2007) A method to distinguish between the de novo induction of thymidine kinase mutants and the selection of pre-existing thymidine kinase mutants in the mouse lymphoma assay. Mutat Res 626:185–190
98. Corsaro CM, Migeon BR (1977) Contact-mediated communication of ouabain resistance in mammalian cells in culture. Nature 268:737–739
99. Dayan J, Deguingand S, Truzman C (1985) Study of the mutagenic activity of 6 hepatotoxic pharmaceutical drugs in the Salmonella typhimurium microsome test, and the HGPRT and Na+/K+ ATPase system in cultured mammalian cells. Mutat Res 157:1–12
100. Moore MM, Harrington-Brock K, Doerr CL, Dearfield KL (1989) Differential mutant quantitation at the mouse lymphoma tk and CHO hgprt loci. Mutagenesis 4:394–403
101. Li AP, Aaron CS, Auletta AE, Dearfield KL, Riddle JC, Slesinski RS, Jr StankowskiLF (1991) An evaluation of the roles of mammalian cell mutationassays in the testing of chemical genotoxicity. Regul Toxicol Pharmacol 14:24–40

102. Puck TT, Waldren CA (1987) Mutation in mammalian cells: theory and implications. Somatic Cell Mol Genet 13:405–409
103. Waldren C, Jones C, Puck TT (1979) Measurement of mutagenesis in mammalian cells. Proc Natl Acad Sci USA 76:1358–1362
104. Ross CD, Lim CU, Fox MH (2005) Assay to measure CD59 mutations in CHO A(L) cells using flow cytometry. Cytometry Part A 66:85–90
105. Zhou H, Xu A, Gillispie JA, Waldren CA, Hei TK (2006) Quantification of CD59-mutants in human-hamster hybrid (AL) cells by flow cytometry. Mutat Res 594:113–119
106. Baumstark-Khan C (2009) A calculation method of PLD recovery based on saturation kinetics. In: Riklis E (ed) Frontiers in radiation biology. VCH & Balaban, Weinheim, pp 513–520
107. Dikomey E, Franzke J (1986) DNA repair kinetics after exposure to X-irradiation and to internal beta-rays in CHO cells. Radiat Environ Biophys 25:189–194
108. Dikomey E, Franzke J (1986) Three classes of DNA strand breaks induced by X-irradiation and internal beta-rays. Int J Radiat Biol 50:893–908
109. Weibezahn KF, Coquerelle T (1981) Radiation induced DNA double strand breaks are rejoined by ligation and recombination processes. Nucleic Acids Res 9:3139–3150
110. Madle S, Dean SW, Andrae U, Brambilla G, Burlinson B, Doolittle DJ, Furihata C, Hertner T, McQueen CA, Mori H (1994) Recommendations for the performance of UDS tests in vitro and in vivo. Mutat Res 312:263–285
111. Mitchell AD, Casciano DA, Meltz ML, San RH, Williams GM, von Halle ES (1983) Unscheduled DNA synthesis tests. A report of the U.S. Environmental protection agency gene-tox program. Mutat Res 123:363–410
112. Efeyan A, Serrano M (2007) p53: guardian of the genome and policeman of the oncogenes. Cell Cycle 6:1006–1010
113. Gomez-Lazaro M, Fernandez-Gomez FJ, Jordán J (2004) p53: twenty five years understanding the mechanism of genome protection. J Physiol Biochem 60:287–307
114. Janus F, Albrechtsen N, Dornreiter I, Wiesmüller L, Grosse F, Deppert W (1999) The dual role model for p53 in maintaining genomic integrity. Cell Mol Life Sci 55:12–27
115. Wu X, Ranganathan V, Weisman DS, Ciccone DN, O'Neill TB, Crick KE, Pierce KA, Lane WS, Rathbun G, Livingston DM, Weaver DT (2000) ATM phosphorylation of Nijmegen breakage syndrome protein is required in a DNA damage response. Nature 405:477–482
116. Lee H, Kwak HJ, Cho IT, Park SH, Lee CH (2009) S1219 residue of 53BP1 is phosphorylated by ATM kinase upon DNA damage and required for proper execution of DNA damage response. Biochem Biophys Res Commun 378:32–36
117. Xu B, O'Donnell AH, Kim ST, Kastan MB (2002) Phosphorylation of serine 1387 in Brca1 is specifically required for the Atm-mediated S-phase checkpoint after ionizing irradiation. Cancer Res 62:4588–4591
118. O'Reilly MA (2005) Redox activation of p21Cip1/WAF1/Sdi1: a multifunctional regulator of cell survival and death. Antioxid Redox Signal 7:108–118
119. Mazumder S, Plesca D, Almasan A (2007) A jekyll and hyde role of cyclin E in the genotoxic stress response: switching from cell cycle control to apoptosis regulation. Cell Cycle 6: 1437–1442
120. Mukherjee B, Kessinger C, Kobayashi J, Chen BP, Chatterjee A, Burma S (2006) DNA-PK phosphorylates histone H2AX during apoptotic DNA fragmentation in mammalian cells. DNA Repair 5:575–590
121. Liebermann DA, Hoffman B (2008) Gadd45 in stress signaling. J Mol Signal 3:15
122. Solier S, Sordet O, Kohn KW, Pommier Y (2009) Death receptor-induced activation of the Chk2- and histone H2AX-associated DNA damage response pathways. Mol Cell Biol 29:68–82
123. Scholzen T, Gerdes J (2000) The Ki-67 protein: from the known and the unknown. J Cell Physiol 182:311–322

124. Wu X, Lieber MR (1996) Protein-protein and protein-DNA interaction regions within the DNA end-binding protein Ku70-Ku86. Mol Cell Biol 16:5186–5193
125. Agrawal A, Yang J, Murphy RF, Agrawal DK (2006) Regulation of the p14ARF-Mdm2–p53 pathway: an overview in breast cancer. Environ Mol Pathol 81:115–122
126. Williams RS, Williams JS, Tainer JA (2007) Mre11-Rad50-Nbs1 is a keystone complex connecting DNA repair machinery, double-strand break signaling, and the chromatin template. Biochem Cell Biol 85:509–520
127. Marcelain K, De La Torre C, González P, Pincheira J (2009) Roles of nibrin and AtM/ATR kinases on the G2 checkpoint under endogenous or radio-induced DNA damage. Biol Res 38:179–185
128. Besson A, Yong VW (2001) Mitogenic signaling and the relationship to cell cycle regulation in astrocytomas. J Neurooncol 51:245–264
129. Essers J, Theil AF, Baldeyron C, van Cappellen WA, Houtsmuller AB, Kanaar R, Vermeulen W (2005) Nuclear dynamics of PCNA in DNA replication and repair. Mol Cell Biol 25:9350–9359
130. Digweed M, Rothe S, Demuth I, Scholz R, Schindler D, Stumm M, Grompe M, Jordan A, Sperling K (2002) Attenuation of the formation of DNA-repair foci containing RAD51 in Fanconi anaemia. Carcinogenesis 23:1121–1126
131. Meek DW (1994) Post-translational modification of p53. Semin Cancer Biol 5:203–210
132. Rocha S, Garrett MD, Campbell KJ, Schumm K, Perkins ND (2005) Regulation of NF-kappaB and p53 through activation of ATR and Chk1 by the ARF tumour suppressor. EMBO J 24:1157–1169
133. Pilch DR, Sedelnikova OA, Redon C, Celeste A, Nussenzweig A, Bonner WM (2003) Characteristics of gamma-H2AX foci at DNA double-strand breaks sites. Biochem Cell Biol 81:123–129
134. Iijima K, Ohara M, Seki R, Tauchi H (2008) Dancing on damaged chromatin: functions of ATM and the RAD50/MRE11/NBS1 complex in cellular responses to DNA damage. J Radiat Res 49:451–464
135. Kang J, Ferguson D, Song H, Bassing C, Eckersdorff M, Alt FW, Xu Y (2009) Functional interaction of H2AX, NBS1, and p53 in ATM-dependent DNA damage responses and tumor suppression. Mol Cell Biol 25:661–670
136. Ward IM, Minn K, Jorda KG, Chen J (2003) Accumulation of checkpoint protein 53BP1 at DNA breaks involves its binding to phosphorylated histone H2AX. J Biol Chem 278:19579–19582
137. Rashi-Elkeles S, Elkon R, Weizman N, Linhart C, Amariglio N, Sternberg G, Rechavi G, Barzilai A, Shamir R, Shiloh Y (2006) Parallel induction of ATM-dependent pro- and antiapoptotic signals in response to ionizing radiation in murine lymphoid tissue. Oncogene 25:1584–1592
138. Feng Z, Hu W, Rajagopal G, Levine AJ (2008) The tumor suppressor p53: cancer and aging. Cell Cycle 7:842–847
139. Gottlieb TM, Oren M (1996) p53 in growth control and neoplasia. Biochim Biophys Acta 1287:77–102
140. Dulic V, Kaufmann WK, Wilson SJ, Tlsty TD, Lees E, Harper JW, Elledge SJ, Reed SI (2009) p53-dependent inhibition of cyclin-dependent kinase activities in human fibroblasts during radiation-induced G1 arrest. Cell 76:1013–1023
141. Karin M, Lin A (2002) NF-kappaB at the crossroads of life and death. Nat Immunol 3:221–227
142. Papa S, Zazzeroni F, Pham CG, Bubici C, Franzoso G (2004) Linking JNK signaling to NF-kappaB: a key to survival. J Cell Sci 117:5197–5208
143. Habraken Y, Piette J (2006) NF-kappaB activation by double-strand breaks. Biochem Pharmacol 72:1132–1141
144. Oberst A, Bender C, Green DR (2008) Living with death: the evolution of the mitochondrial pathway of apoptosis in animals. Cell Death Different 15:1139–1146

145. Abu-Qare AW, Abou-Donia MB (2001) Biomarkers of apoptosis: release of cytochrome c, activation of caspase-3, induction of 8-hydroxy-2′-deoxyguanosine, increased 3-nitrotyrosine, and alteration of p53 gene. J Toxicol Environ Health Part B Crit Rev 4:313–332
146. Bradbury DA, Simmons TD, Slater KJ, Crouch SP (2000) Measurement of the ADP:ATP ratio in human leukaemic cell lines can be used as an indicator of cell viability, necrosis and apoptosis. J Immunol Methods 240:79–92
147. Sundberg SA (2000) High-throughput and ultra-high-throughput screening: solution- and cell-based approaches. Curr Opin Biotechnol 11:47–53
148. Fischbach M, Bromley P (2001) Recombinant cell lines for stress reporter assays. Cell Biol Toxicol 17:335–351
149. King AV, Jones PA (2003) In-house assessment of a modified in vitro cytotoxicity assay for higher throughput estimation of acute toxicity. Toxicol In Vitro 17:717–722
150. Alley MC, Scudiero DA, Monks A, Hursey ML, Czerwinski MJ, Fine DL, Abbott BJ, Mayo JG, Shoemaker RH, Boyd MR (1988) Feasibility of drug screening with panels of human tumor cell lines using a microculture tetrazolium assay. Cancer Res 48:589–601
151. Riss TL, Moravec RA (2004) Use of multiple assay endpoints to investigate the effects of incubation time, dose of toxin, and plating density in cell-based cytotoxicity assays. Assay Drug Dev Technol 2:51–62
152. Jost LM, Kirkwood JM, Whiteside TL (1992) Improved short- and long-term XTT-based colorimetric cellular cytotoxicity assay for melanoma and other tumor cells. J Immunol Methods 147:153–165
153. Fan F, Wood KV (2007) Bioluminescent assays for high-throughput screening. Assay Drug Dev Technol 5:127–136
154. Wesierska-Gadek J, Gueorguieva M, Ranftler C, Zerza-Schnitzhofer G (2005) A new multiplex assay allowing simultaneous detection of the inhibition of cell proliferation and induction of cell death. J Cell Biochem 96:1–7
155. O'Brien MA, Moravec RA, Riss TL, Bulleit RF (2008) Homogeneous, bioluminescent proteasome assays. Methods Mol Biol 414:163–181
156. Baumstark-Khan C, Horneck G (2007) Results from the "Technical workshop on genotoxicity biosensing" on the microscale fluorometric assay of DNA-unwinding. Anal Chim Acta 593:75–81
157. Ullmann K, Müller C, Steinberg P (2008) Two essential modifications strongly improve the performance of the Fast Micromethod to identify DNA single- and double-strand breaks. Arch Toxicol 82:861–867
158. Frieauff W, Hartmann A, Suter W (2001) Automatic analysis of slides processed in the comet assay. Mutagenesis 15:133–137
159. Kiskinis E, Suter W, Hartmann A (2002) High throughput comet assay using 96-well plates. Mutagenesis 17:37–43
160. Ashcroft R, Lopez P (2000) Commercial high speed machines opennew opportunities in high throughput flow cytometry (HTFC). J Immunol Methods 243:13–24
161. Forand A, Dutrillaux B, Bernardino-Sgherri J (2004) Gamma-H2AX expression pattern in non-irradiated neonatal mouse germ cells and after low-dose gamma-radiation: relationships between chromatid breaks and DNA double-strand breaks. Biol Reprod 71:643–649
162. Abraham VC, Taylor DL, Haskins JR (2004) High content screening applied to large-scale cell biology. Trends Biotechnol 22:15–22
163. Cervinka M, Cervinkova Z, Rudolf E (2008) The role of time-lapse fluorescent microscopy in the characterization of toxic effects in cell populations cultivated in vitro. Toxicol In Vitro 22:1382–1386
164. Evans JG, Matsudaira P (2007) Linking microscopy and high content screening in large-scale biomedical research. Methods Mol Biol 356:33–38
165. Rausch O (2006) High content cellular screening. Curr Opin Chem Biol 10:316–329
166. Diaz D, Scott A, Carmichael P, Shi W, Costales C (2007) Evaluation of an automated in vitro micronucleus assay in CHO-K1 cells. Mutat Res 630:1–13

167. He YD (2006) Genomic approach to biomarker identification and its recent applications. Cancer Biomarkers 2:103–133
168. Leighton JK (2005) Application of emerging technologies in toxicology and safety assessment: regulatory perspectives. Int J Toxicol 24:153–155
169. Aardema MJ, MacGregor JT (2002) Toxicology and genetic toxicology in the new era of "toxicogenomics": impact of "-omics" technologies. Mutat Res 499:13–25
170. Waters MD, Fostel JM (2004) Toxicogenomics and systems toxicology: aims and prospects. Nat Rev Genet 5:936–948
171. Xu ZL, Mizuguchi H, Ishii-Watabe A, Uchida E, Mayumi T, Hayakawa T (2001) Optimization of transcriptional regulatory elements for constructing plasmid vectors. Gene 272: 149–156
172. López-Maury L, Marguerat S, Bähler J (2008) Tuning gene expression to changing environments: from rapid responses to evolutionary adaptation. Nat Rev Genet 9:583–593
173. Edelman GM, Meech R, Owens GC, Jones FS (2000) Synthetic promoter elements obtained by nucleotide sequence variation and selection for activity. Proc Nat Acad Sci USA 97:3038–3043
174. Alcaraz-Pérez F, Mulero V, Cayuela ML (2008) Application of the dual-luciferase reporter assay to the analysis of promoter activity in Zebrafish embryos. BMC Biotechnol 8:81
175. Arnone MI, Dmochowski IJ, Gache C (2004) Using reporter genes to study cis-regulatory elements. Methods Cell Biol 74:652
176. Schlaeger EJ, Kitas EA, Dorn A (2003) SEAP expression in transiently transfected mammalian cells grown in serum-free suspension culture. Cytotechnology 42:47–55
177. Michelini E, Cevenini L, Mezzanotte L, Ablamsky D, Southworth T, Branchini BR, Roda A (2008) Combining intracellular and secreted bioluminescent reporter proteins for multicolor cell-based assays. Photochem Photobiol Sci 7:212–217
178. Tannous BA, Kim DE, Fernandez JL, Weissleder R, Breakefield XO (2005) Codon-optimized Gaussia luciferase cDNA for mammalian gene expression in culture and in vivo. Mol Ther 11:435–443
179. Baumstark-Khan C, Hellweg CE, Horneck G (2002) On the suitability of red and enhanced green fluorescent protein (DsRed/EGFP) as reporter combination. In: Hoffmann K-H (ed), Springer, Berlin, pp 1–13
180. Hellweg CE, Baumstark-Khan C, Rettberg P, Horneck G (2001) The suitability of enhanced green fluorescent protein (EGFP) as reporter component for bioassays. Anal Chim Acta 426:175–184
181. Shaner NC, Steinbach PA, Tsien RY (2005) A guide to choosing fluorescent proteins. Nat Methods 2:905–909
182. Jeng MH, Shupnik MA, Bender TP, Westin EH, Bandyopadhyay D, Kumar R, Masamura S, Santen RJ (1998) Estrogen receptor expression and function in long-term estrogen-deprived human breast cancer cells. Endocrinology 139:4164–4174
183. Donato MT, Lahoz A, Castell JV, Gómez-Lechón MJ (2008) Cell lines: a tool for in vitro drug metabolism studies. Curr Drug Metab 9:1–11
184. Scheuenpflug J, Krebsfänger N, Doehmer J (2005) Heterologous co-expression of human cytochrome P450 1A2 and polymorphic forms of N-acetyltransferase 2 for studies on aromatic amines in V79 Chinese hamster cells. Altern Lab Anim 33:561–577
185. Zhuge J, Luo Y, Yu YN (2003) Heterologous expression of human cytochrome P450 2E1 in HepG2 cell line. World J Gastroenterol 9:2732–2736
186. Guillouzo A, Corlu A, Aninat C, Glaise D, Morel F, Guguen-Guillouzo C (2007) The human hepatoma HepaRG cells: a highly differentiated model for studies of liver metabolism and toxicity of xenobiotics. Chem-Biol Interact 168:66–73
187. Chen JJ (2007) Key aspects of analyzing microarray gene-expression data. Pharmacogenomics 8:473–482
188. Lettieri T (2006) Recent applications of DNA microarray technology to toxicology and ecotoxicology. Environ Health Perspect 114:4–9

189. Coombe DR, Nakhoul AM, Stevenson SM, Peroni SE, Sanderson CJ (1998) Expressed luciferase viability assay (ELVA) for the measurement of cell growth and viability. J Immunol Methods 215:145–150
190. Elliott G, McGrath J, Crockett-Torabi E (2000) Green fluorescent protein: a novel viability assay for cryobiological applications. Cryobiology 40:360–369
191. Baumstark-Khan C, Hellweg CE, Palm M, Horneck G (2001) Enhanced green fluorescent protein (EGFP) for space radiation research using mammalian cells in the International Space Station. Phys Med 17:210–214
192. Hellweg CE, Baumstark-Khan C, Horneck G (2001) Enhanced green fluorescent protein as reporter protein for biomonitoring of cytotoxic effects in mammalian cells. Anal Chim Acta 427:191–199
193. Hellweg CE, Arenz A, Baumstark-Khan C (2007) Assessment of space environmental factors by cytotoxicity bioassays. Acta Astronaut 60:525–533
194. Quiñones A, Rainov NG (2001) Identification of genotoxic stress in human cells by fluorescent monitoring of p53 expression. Mutat Res 494:73–85
195. Gorbunov NV, Morris JE, Greenberger JS, Thrall BD (2002) Establishment of a novel clonal murine bone marrow stromal cell line for assessment of p53 responses to genotoxic stress. Toxicology 179:257–266
196. Ohno K, Ishihata K, Tanaka-Azuma Y, Yamada T (2008) A genotoxicity test system based on p53R2 gene expression in human cells: assessment of its reactivity to various classes of genotoxic chemicals. Mutat Res 656:27–35
197. Ohno K, Tanaka-Azuma Y, Yoneda Y, Yamada T (2005) Genotoxicity test system based on p53R2 gene expression in human cells: examination with 80 chemicals. Mutat Res 588: 47–57
198. Arenz A (2007) Design and validation of a bioassay for detection of induced gene activity to be used at the International Space Station, Faculty of Mathematics and Natural Sciences, University of Cologne, Germany
199. Hellweg CE, Baumstark-Khan C, Horneck G (2003) Generation of stably transfected mammalian cell lines as fluorescent screening assay for NF-kappaB activation-dependent gene expression. J Biomol Screening 8:511–521
200. Hellweg CE, Arenz A, Bogner S, Schmitz C, Baumstark-Khan C (2006) Activation of nuclear factor kappa B by different agents: influence of culture conditions in a cell-based assay. Ann NY Acad Sci 1091:191–204
201. Baumstark-Khan C, Hellweg CE, Arenz A, Meier MM (2005) Cellular monitoring of the nuclear factor kappaB pathway for assessment of space environmental radiation. Radiat Res 1664:527–530
202. Hellweg CE, Baumstark-Khan C (2007) Detection of UV-induced activation of NF-kappaB in a recombinant human cell line by means of enhanced green fluorescent protein (EGFP). Radiat Environ Biophys 46:269–279
203. Todd MD, Lee MJ, Williams JL, Nalezny JM, Gee P, Benjamin MB, Farr SB (1995) Stress responses to DNA damaging agents in the human colon carcinoma cell line RKO. Fundam Appl Toxicol 28:118–128
204. Hastwell PW, Chai LL, Roberts KJ, Webster TW, Harvey JS, Rees RW, Walmsley RM (2006) High-specificity and high-sensitivity genotoxicity assessment in a human cell line: validation of the GreenScreen HC GADD45a-GFP genotoxicity assay. Mutat Res 607:160–175
205. Walmsley RM (2008) GADD45a-GFP GreenScreen HC genotoxicity screening assay. Expert Opin Drug Metab Toxicol 4:827–835
206. Beard SE, Capaldi SR, Gee P (2009) Stress responses to DNA damaging agents in the human colon carcinoma cell line RKO. Mutat Res 371:1–13
207. O'Connell-Rodwell CE, Shriver D, Simanovskii DM, McClure C, Cao YA, Zhang W, Bachmann MH, Beckham JT, Jansen ED, Palanker D, Schwettman HA, Contag CH (2004) A genetic reporter of thermal stress defines physiologic zones over a defined temperature range. FASEB J 18:264–271

208. Fischbach M, Sabbioni E, Bromley P (1993) Induction of the human growth hormone gene placed under human hsp70 promoter control in mouse cells: a quantitative indicator of metal toxicity. Cell Biol Toxicol 9:177–188
209. Yoshizumi M, Wang H, Hsieh CM, Sibinga NE, Perrella MA, Lee ME (1997) Down-regulation of the cyclin A promoter by transforming growth factor-beta1 is associated with a reduction in phosphorylated activating transcription factor-1 and cyclic AMP-responsive element-binding protein. J Biol Chem 272:22259–22264
210. Imreh G, Beckman M, Iverfeldt K, Hallberg E (1998) Noninvasive monitoring of apoptosis versus necrosis in a neuroblastoma cell line expressing a nuclear pore protein tagged with the green fluorescent protein. Exp Cell Res 238:371–376
211. Mahajan NP, Harrison-Shostak DC, Michaux J, Herman B (1999) Novel mutant green fluorescent protein protease substrates reveal the activation of specific caspases during apoptosis. Chem Biol 6:401–409
212. Strebel A, Harr T, Bachmann F, Wernli M, Erb P (2001) Green fluorescent protein as a novel tool to measure apoptosis and necrosis. Cytometry 43:126–133
213. Johnson JM, Latimer JJ (2005) Analysis of DNA repair using transfection-based host cell reactivation. Methods Mol Biol 291:321–335
214. Stoklosa T, Poplawski T, Koptyra M, Nieborowska-Skorska M, Basak G, Slupianek A, Rayevskaya M, Seferynska I, Herrera L, Blasiak J, Skorski T (2008) BCR/ABL inhibits mismatch repair to protect from apoptosis and induce point mutations. Cancer Res 68:2576–2580
215. Zhou B, Huang C, Yang J, Lu J, Dong Q, Sun LZ (2009) Preparation of heteroduplex EGFP plasmid for in vivo mismatch repair activity assay. Anal Biochem 388:167–169
216. Roguev A, Russev G (2000) Two-wavelength fluorescence assay for DNA repair. Anal Biochem 287:313–318
217. Collis SJ, Sangar VK, Tighe A, Roberts SA, Clarke NW, Hendry JH, Margison GP (2002) Development of a novel rapid assay to assess the fidelity of DNA double-strand-break repair in human tumour cells. Nucleic Acids Res 30:e1
218. Slebos RJ, Taylor JA (2001) A novel host cell reactivation assay to assess homologous recombination capacity in human cancer cell lines. Biochem Biophys Res Commun 281:212–219

Adv Biochem Engin/Biotechnol (2010) 118: 153–164
DOI: 10.1007/10_2009_4

Published online: 21 May 2009

Live Cell Optical Sensing for High Throughput Applications

Ye Fang

Abstract Live cell optical sensing employs label-free optical biosensors to non-invasively measure stimulus-induced dynamic mass redistribution (DMR) in live cells within the sensing volume of the biosensor. The resultant DMR signal is an integrated cellular response, and reflects cell signaling mediated through the cellular target(s) with which the stimulus intervenes. This article describes the uses of live cell optical sensing for probing cell biology and ligand pharmacology, with an emphasis of resonant waveguide grating biosensor cellular assays for high throughput applications.

Keywords Cell-based assay • Cell signaling • Dynamic mass redistribution • G protein-coupled receptor • High throughput screening • Optical biosensor • Resonant waveguide grating biosensor

Contents

Y. Fang
Biochemical Department, Science and Technology Division,
Corning Incorporated, Corning, New York 14831
e-mail: fangy2@corning.com

Abbreviations

β_2AR	β_2 Adrenergic receptor
GPCR	G protein-coupled receptor
H_1R	Histamine receptor subtype 1
HTS	High throughput screening
RWG	Resonant waveguide grating
SPR	Surface plasmon resonance

1 Introduction

Cell-based assays that examine the activities of live cells have become the powerhouse in both cell biology and drug discovery in the past decades, due to their ability to facilitate the measurements of mode of action, pathway activation, toxicity, and phenotypic cellular responses of exogenous stimuli. However, most cell-based assays suffer from two drawbacks. First, these assays typically measure "point-of-contacts", such as the generation of a second messenger, the translocation of a tagged protein, or the expression of a reporter gene, in a myriad of cellular events downstream ligand-induced receptor activation [1]. Thus, the activity of a ligand obtained is only related to the specific cellular event measured, but does not reflect the full potentials of the ligand acting on the live cell or cell systems [2, 3]. Second, these assays require certain manipulations of cellular contexts or background (e.g., over-expression of targets with or without a readout tag); and such manipulations could result in false information of receptor biology [4, 5].

Optical biosensors including surface plasmon resonance (SPR), resonant waveguide grating (RWG) and plasmon-waveguide resonance are routinely used for biomolecular interaction analysis [5, 6]. These biosensors are label-free and are capable of measuring minute changes in local index of refraction at/near the biosensor surface. Recently, we [7–13] and others [14, 15] had applied these biosensors for live cell sensing. We have found that these biosensors enable monitoring endogenous receptor activation in real time, leading to high-information and physiologically relevant measures of a receptor–ligand pair [see reviews 7, 8]. These assays do not require prior knowledge of cell signaling, and are pathway-unbiased [9, 10]. Instead of measuring single point-of-contacts within complex signaling cascades that most cellular assays do, the live cell optical sensing relies on an integrated cellular response relating to dynamic mass redistribution (DMR) to probe receptor biology and ligand pharmacology [11–13]. This article reviews the recent advances in high throughput instrumentation, and theoretical considerations and applications of live cell optical sensing.

2 High Throughput Biosensor Systems

RWG belongs to a family of optical waveguide-based biosensors, and uses the resonant coupling of light into a waveguide by means of a diffraction grating (Fig. 1). A broadband and polarized light is used to illuminate the waveguide; light of a specific wavelength (the resonant wavelength) at which a maximal incoupling efficiency is achieved is coupled into and propagate along the waveguide [16]. The resonant light eventually leaks out the waveguide film and reflects back. The change in resonant wavelength is in proportion to change in local mass density at the sensor surface.

A significant advantage of RWG over conventional SPR is that given appropriate biosensor design, lights at nominally normal incident angle can be used to illuminate the biosensor [17]. This is important for sampling large numbers of biosensors simultaneously, which is a prerequisite for high throughput (HT) screening. Assaying samples in a typical HTS often directly takes place in the Society for Biomolecular Sciences standard microtiter plates, such as 384well microplates. The Epic® system from Corning Inc is the first system that is suitable for both biochemical- and cell-based assays in high throughput. This system consists of a RWG detector, an external liquid handling accessory and a scheduler, such that it can process large numbers of microplates using kinetic measures for high information content screening [10], or using end-point measures for HT screening [17].

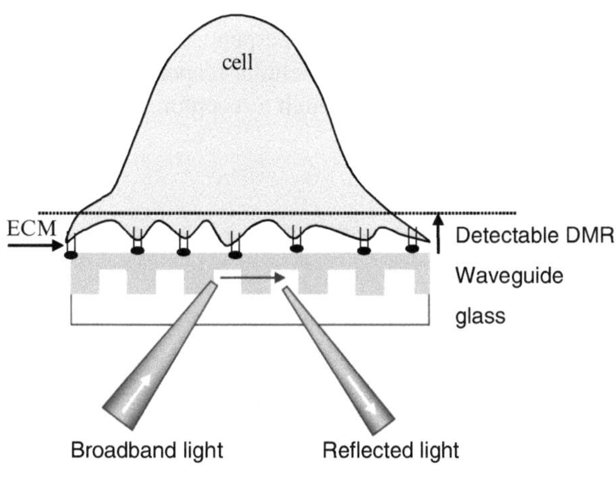

Fig. 1 Principle of live cell RWG sensing. Cells are brought to contact with a biosensor surface through interaction with the extracellular matrix (ECM).

3 Live Cell Optical Sensing Detects an Integrated Cellular Response

Live cell optical sensing generally exploits a surface bound evanescent wave [16, 18] to characterize stimulus-induced cellular responses [9–14]. For RWG biosensors, the evanescent wave is an electromagnetic field created by the total internal reflection of resonant light in the waveguide thin film, whose intensity exponentially decays away from the sensor surface. The distance from the sensor surface at which the electromagnetic field strength has decreased to $1/e$ of its initial value is the penetration depth or sensing volume (typically ~150nm). Considering the large dimension of live cells, a live cell RWG sensing system can be viewed as a three-layer waveguide: a substrate with an optical grating, a high index of refraction waveguide coating, and a cell layer. Since the local index of refraction within a cell is a function of density and its distribution of biomass (e.g., proteins, molecular complexes) [19], a ligand-induced change in effective refractive index (i.e., the detected signal) is believed to be governed by [13]:

$$\Delta N = S(C)\alpha d \sum_i \Delta C_i(t)\left[e^{\frac{-z_i}{\Delta Z_C}} - e^{\frac{-z_{+1i}}{\Delta Z_C}} \right],$$

where $S(C)$ is the sensitivity of the biosensor to the cell layer, ΔZ_c the penetration depth into the cell layer, α the specific refraction increment (about 0.18/mL/g for proteins), $\Delta C_i(t)$ the change in local concentration of biomolecules at the given location at a specific time, z_i the distance where the mass redistribution occurs, and d an imaginary thickness of a slice within the cell layer. This theory predicates, as confirmed by experimental studies [9–13], that the biosensor is capable of non-invasively monitoring cell signaling in real time, a stimulus-induced shift in resonant wavelength is dominated by the DMR perpendicular to the sensor surface, and the resultant DMR signal is an integrated cellular response containing contributions from many cellular events mediated through a receptor.

4 Live Cell Optical Sensing for Receptor Biology

Live cell optical sensing has found broad applications in cell biology, including cell signaling [7–14, 20, 21]. Cell signaling is part of a complex system of communication that governs basic cellular activities and coordinates cell actions. Cell signaling is encoded by spatial gradients and temporal modifications of cellular molecules upon receiving an exogenous signal [22–24]. The temporal modifications of key proteins in the posttranslational state, enzymatic activity, or total level can act as molecular signals that are relayed and interpreted to control cell function [25, 26]. Furthermore, the spatial targeting of signaling molecules and assemblies to appropriate sites, coupled with its temporal dynamics, is crucial to regulate the specificity and efficiency of protein–protein interactions, and to spatially separate protein activation and deactivation mechanisms, thus enabling the precise control of the amplitude, duration and kinetics of cell signaling [27]. As a result, there is often dynamic relocation of cellular matters within the cells, leading to a dynamic, directional and directed mass redistribution. A DMR when occurring in the sensing volume of a biosensor can be detected non-invasively by a label-free optical biosensor [11, 12].

Because of the integrative nature of DMR responses, live cell optical sensing is able to examine the complexity of receptor biology. An example is the signaling of endogenous bradykinin B_2 receptor in a human skin cancerous cell line A431. Bradykinin, a vasoactive nonapeptide, is implicated in the regulation of many physiological and pathophysiological responses, including vascular permeability, inflammation, pain, and neurotransmitter release [28]. We found, using live cell optical sensing, that the signaling of B_2 receptor in A431 is dependent on cellular contexts [12, 17]. In proliferating A431 cells, low doses (<100 nM) of bradykinin trigger G_s-mediated signaling, while high doses (>100 nM) of bradykinin favors G_q-signaling. On the other hand, bradykinin between 0.5nM and about 100nM mediated dual signaling – G_s and G_q-pathways – in partially quiescent A431 cells obtained through continuous culturing with 0.1% fetal bovine serum for about 20hours (Fig. 2) [12]. We also found that in the partially quiescent cells the bradykinin response is specific to B_2 receptor activation, and is sensitive to the integrity of lipid rafts. Using chemical biology approach which exploits chemical intervention to probe receptor biology, not only critical downstream cascades

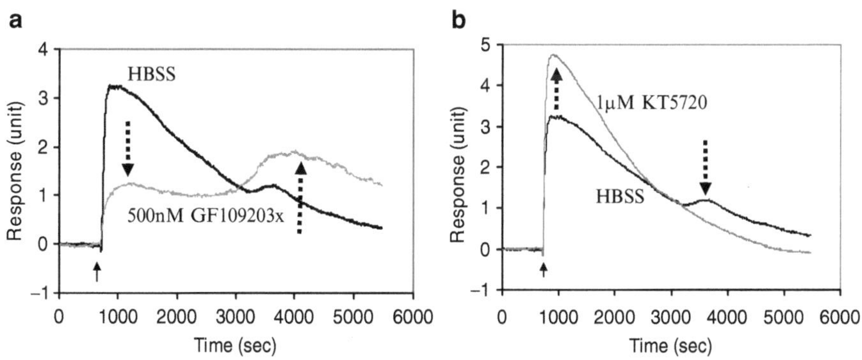

Fig. 2 Dual signaling of bradykinin B_2 receptor in A431 cells. The bradykinin-induced DMR responses of cells pre-treated with the "vehicle" (HBSS) were in comparison with those pretreated with 500nM GF109203x (a protein kinase C inhibitor) (**a**), or 1μM KT5720 (a protein kinase A inhibitor) (**b**). The bradykinin concentration was 16nM. The *solid arrows* indicate the time when the agonist solution is introduced. 1unit in the *y*-axis equals to 100 pm shift in resonant wavelength. Chemical modulation showed that in the quiescent cells blockage of protein kinase C activity partially attenuates the early response but potentiates the late response (**a**), while blockage of protein kinase A activity potentiates the early response but attenuates the late response (**b**). These results suggest that the activation of B_2 receptors leads to both G_q and G_s signaling, and the two signaling pathways counter-regulate each other. Data was reproduced with permission from [12].

including protein kinase C, protein kinase A, and epidermal growth factor receptor, but also several cellular processes including endocytosis and cytoskeleton modulation had been found to be involved in B_2 signaling. These findings are in great agreement with previous studies using conventional cell biology approaches, which showed that the B_2 receptor in A431 can be mitogenic or anti-mitogenic, depending on the cellular states and the doses of bradykinin [29, 30].

5 Live Cell Optical Sensing for Ligand Pharmacology

Ligand pharmacology can be complicated. This is partly due to the intrinsic property of a ligand that often displays off-target effects and/or pathway-biased efficacies [2, 31], and partly due to the complexity of receptor signaling [32–36] . The pathway-biased efficacy is also referred to ligand-directed functional selectivity. A classic example is the G protein-coupled receptor (GPCR) signaling. A GPCR may couple simultaneously to more than one G protein subtype, and interact with other signaling molecules such as arrestins [33, 34]. In many cases the activation of a receptor can mediate both G protein-dependent and independent signaling, often in a ligand-dependent manner [35, 36]. As a result, GPCRs display rich behaviors in cells [37], and many ligands can induce operative bias to favor specific portions of the cell machinery and exhibit pathway-biased efficacies [31, 2]. It is obviously difficult in practice for conventional and pathway-biased cellular assays to systematically determine the signaling capacity of GPCR ligands.

Since live cell optical sensing is a highly sensitive functional assay and is applicable to wide arrays of targets and cellular processes [7–14], it can be used to study the systems cell pharmacology including ligand-directed functional selectivity of a drug molecule acting on cells. The high sensitivity enables detecting weak biological signals such as those provided by ligands of low efficacy and/or affinity. The pathway unbiased nature, coupled with the real time kinetics, allows the direct detection of possible pathway-biased activity of ligands.

Figure 3 shows the structure, DMR signatures and parameters of four agonists acting through endogenous β_2AR in A431 using live cell RWG sensing. The four agonists differ greatly in their abilities to activate the β_2AR, thus triggering signaling [38, 39]. The functional selectivity of agonists examined is clearly evident in their DMR characteristics. Interestingly, multi-parameter analysis uncovers distinct DMR patterns linking the structure to activities of these ligands. All four agonists led to similar responses but with great difference in fine features. Both the amplitudes of P- (positive-) and N- (negative-) DMR events reflect the efficacy of ligands. The full agonist epinephrine led to the greatest P- and N-DMR events. The occurrence of N-DMR is believed to be an indicator for the ability of the agonists to elevate intracellular cAMP level. The very weak partial agonist catechol was unable to produce any N-DMR event at all. The P-DMR kinetics, as shown by its $t_{1/2}$, is believed to be an effective indicator for ligands to induce receptor internalization. Both dopamine and catechol that are ineffective in causing receptor internalization led to faster P-DMR, compared to epinephrine or norepinephrine. Similarly, the transition time τ for the

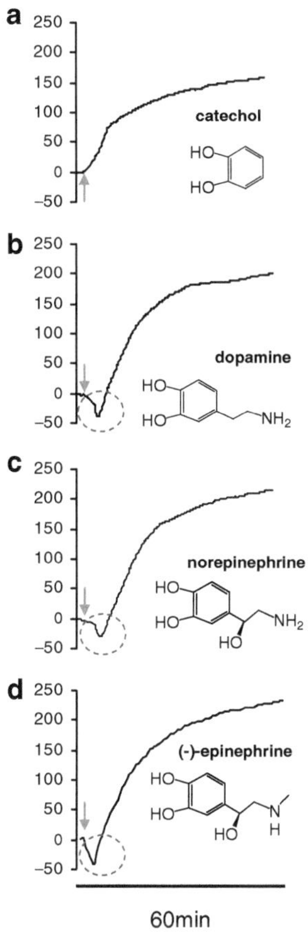

P-DMR(pm)	N-DMR(pm)	τ(s)	$t_{1/2}$(s)
152±13	0±3	130±15	289±42
214±32	31±4	252±15	480±23
209±16	29±5	180±17	559±15
232±12	37±7	132±20	521±25

Fig. 3 The structures of β_2AR ligands, and their DMR signals and parameters in quiescent A431 cells. The ligands included the full agonist (–)epinephrine (8nM), the strong partial agonist norepinephrine (100nM), the partial agonist dopamine (32μM), and the very weak partial agonist catechol (500μM). The *grey arrows* indicated the time when the agonist is introduced. The *y*-axis is the response in picometer (pm). Data was reproduced with permission from [10].

P-DMR to occur is also agonist-dependent. These results illustrate the power of live cell optical sensing for probing ligand-directed functional selectivity.

6 Live Cell Optical Sensing for High Throughput Screening

HT screening that quickly screen massive numbers of compounds is crucial to drug discovery and development, due to the increasing size of compound library and the increasing numbers of druggable targets. Live cell optical sensing holds great potentials in HTS. Unlike conventional assays that often lead to potential false negatives due to the pathway biased nature, as well as high false positives due to labels or cell engineering, live cell optical sensing offers a non-invasive and manipulation-free alternative to assay endogenous cellular targets such as GPCRs in native cells. Additional advantages of live cell optical sensing for HTS include the ability of multiplexing [1, 7], and of multi-modes [40].

An end-point HTS assay using Epic® system has been carried out for the endogenous β_2-adrenergic receptor (β_2AR) in A431. Results showed that the time to assay a single 384well plate is short (~5min per plate), and the assay is quite robust with a Z' of greater than 0.7 [17]. This study has stimulated high interests in industrial companies for potential HTS applications. Several successful HTS campaigns using live cell optical sensing have been performed recently by pharmaceutical companies.

We have recently developed a duplexed label-free optical biosensor cellular assay for simultaneously assaying two endogenous receptors, the G_q-coupled histamine receptor (H_1R) and the G_s-coupled β_2AR, in A431 cells [40]. Coupled with both agonism and antagonism modes in a single kinetic assay, we have shown that the agonist screening not only identified all full agonists for both H_1- and β_2-receptors, but also detected pathway-biased ligands for the β_2AR (Fig. 4). Furthermore, the succeeding antagonist screening documented all known antagonists in the library for either H_1 or β_2 receptors. This is the first demonstration of a single cellular assay that is capable of screening ligands against two GPCRs coupled to distinct G proteins, and highlights the power of pathway-unbiased and label-free biosensor cellular assays for GPCR screens.

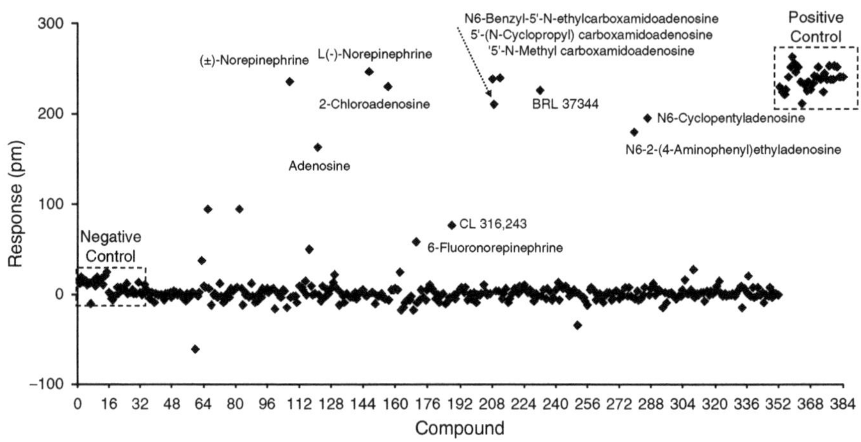

Fig. 4 An example of duplexed receptor screening in a 384well microplate format. The responses of quiescent A431 cells, in the initial agonism mode, were plotted as a function of compound. The quiescent A431 cells were stimulated with a subset of LOPAC library of compounds. The shift in resonant wavelength before and 50min after stimulation was calculated for each well. The negative controls (the "vehicle") led to a response of 8 ± 7pm ($n = 32$), while the positive controls (2nM (–)-epinephrine) led to a response of 230 ± 12pm ($n = 32$), indicating a robust assay with a Z′ factor of 0.74. This end-point screening identified all agonists in this subset of compounds for both the endogenous β_2AR and adenosine receptors in A431. Several other compounds whose mechanisms to trigger the DMR responses are unknown were also identified. Data was reproduced with permission from [40]

7 Conclusion

The drug discovery process has been constantly evolving from affinity- and targeted-based screens to systems biology- or biological or clinical activity-based screens. The advent of optical biosensors, particularly high throughput instruments and cell-based assays, will accelerate acceptance of optical biosensors in many points at many stages of the drug discovery process, for which both high information content and high throughput are important.

Acknowledgement The author thanks Ann M. Ferrie, Elizabeth Tran, Florence Verrier, Haiyan Sun, Huayun Deng, Vasiliy Goral, Chris Wu, Norman Fontaine, Guangshan Li, Meenal Soni, Deepti Mudaliar, Jun Xi, Mala Johnson, Geetha Bhagavatula, Colin Eichinger, Roger Wang, Dana Gary, Aneesh Badrinarayan, Gordon Shedd, Terry Goodrich, John Peanasky, David Randle, Alice Gao, Jeff Scibek, Joydeep Lahiri, Anthony Frutos, David Pastel, and Ron Verkleeren for their past and current contributions in the development of live cell optical sensing.

References

1. Fang Y, Frutos AG, Verkleeren R (2008) Label-free cell-based assays for GPCR screening. Combinatorial Chemistry High Throughput Screening 11: 357–369
2. Galandrin S, Oligny-Longpre G, Bouvier M (2007) The evasive nature of drug efficacy: implications for drug discovery. Trends in Pharmacol Sci 8:423–430
3. Gribbon P, Sewing A (2003) Fluorescence readouts in HTS: no gain without pain? Drug Discovery Today 8:1035–1043
4. Balla T, Varnai P (2002) Visualizing cellular phosphoinositide pools with GFP-fused protein modules. Science STKE 125: 13
5. Cooper MA (2006) Optical biosensors: where next and how soon. Drug Discovery Today 11: 1061–1067
6. Rich RL, Myszka DG (2007) Higher-throughput, label-free, real-time molecular interaction analysis. Anal Biochem 361:1–6
7. Fang Y (2006) Label-free cell-based assays with optical biosensors in drug discovery. Assay Drug Dev Technol 4:583–595
8. Fang Y (2007) Non-invasive optical biosensor for probing cell signaling. Sensors 7:2316–2329
9. Fang Y, Li G, Ferrie AM (2007) Non-invasive optical biosensor for assaying endogenous G protein-coupled receptors in adherent cells. J Pharmacol Toxicol Methods 55:314–322
10. Fang Y, Ferrie AM (2008) Label-free optical biosensor for ligand-directed functional selectivity acting on β2 adrenoceptor in living cells. FEBS Lett 582:558–564
11. Fang Y, Ferrie AM, Fontaine NH, Yuen PK (2005) Characteristics of dynamic mass redistribution of EGF receptor signaling in living cells measured with label free optical biosensors. Anal Chem 77:5720–5725
12. Fang Y, Li G, Peng J (2005) Optical biosensor provides insights for bradykinin B2 receptor signaling in A431 cells. FEBS Lett 579:6365–6374
13. Fang Y, Ferrie AM, Fontaine NH, Mauro J, Balakrishnan J (2006) Resonant waveguide grating biosensor for living cell sensing. Biophys J 91:1925–1940
14. Tanaka M, Hiragun T, Tsutsui T, Yanase Y, Suzuki H, Hide M (2008) Surface plasmon resonance biosensor detects the downstream events of active PKCβ in antigen-stimulated mast cells. Biosens Bioelectron 23:1652–1658
15. Horath R, Pedersen HC, Skivesen N, Selmeczi D, Larsen NB (2005) Monitoring of living cell attachment and spreading using reverse symmetry waveguide sensing. Appl Phys Lett 86: e071101

16. Tiefenthaler K, Lukosz W (1989) Sensitivity of grating couplers as integrated-optical chemical sensors. J Opt Soc Am B 6:209–220
17. Li G, Ferrie AM, Fang Y (2006) Label-free profiling of endogenous G protein-coupled receptors using a cell-based high throughput screening technology. J Assoc Lab Automat 11:181–187
18. Ritchie RH (1957) Plasma losses by fast electrons in thin films. Phys Rev 106:874–881
19. Barer R, Joseph S (1954) Refractometry of living cells. Part I. Basic principles. Quart J Microsc Science 95:399–423
20. Lee PH, Gao A, van Staden C, Ly J, Salon J, Xu A, Fang Y, Verkleeren R (2008) Evaluation of dynamic mass redistribution technology for pharmacological studies of recombinant and endogenously expressed G protein-coupled receptors. Assay Drug Develop Technol 6:83–94
21. Schroeder R, Merten N, Mathiesen JM, Martini L, Kruljac-Letunic A, Krop F, Blaukat A, Fang Y, Tran E, Ulven T, Drewke C, Whistler J, Pardo L, Gomeza J, Kostenis E (2009) The C-terminal tail of CRTH2 is a key molecular determinant that constrains Gαi- and downstream-signaling cascade activation. J Biol Chem. 284:1324–1336
22. Zaccolo M, Pozzan T (2002) Discrete microdomains with high concentration of cAMP in stimulated rat neonatal cardiac myocytes. Science 295:1711–1715
23. Kholodenko BN (2003) Four-dimensional organization of protein kinase signaling cascades: the roles of diffusion, endocytosis and molecular motors. J Exper Biol 206:2073–2082
24. Bonneau R (2008) Learning biological networks: from modules to dynamics. Nat Chem Biol 4:658–664
25. Morris AJ, Malbon CC (1999) Physiological regulation of G protein-linked signaling. Physiol Rev 79:1373–1430
26. Aldridge BB, Burke JM, Lauffenburger DA, Sorger PK (2006) Physicochemical modelling of cell signalling pathways. Nat Cell Biol 8:1195–1203
27. Sato TK, Overduin M, Emr SD (2001) Location, location, location: membrane targeting directed by PX domains. Science 294:1881–1885
28. Marceau F, Regoli D (2004) Bradykinin receptor ligands: therapeutic perspectives. Nat Rev Drug Discov 3:845–852
29. Liebmann C, Graness A, Ludwig B, Adomeit A, Boehmer A, Boehmer F-D, Nürnberg B, Wetzker R (1996) Dual bradykinin B2 receptor signalling in A431 human epidermoid carcinoma cells: activation of protein kinase C is counteracted by a Gs-mediated stimulation of the cyclic AMP pathway. Biochem J 313:109–118
30. Graness A, Hanke S, Boehmer F-D, Presek P, Liebmann C (2000) Protein-tyrosine-phosphatase-mediated epidermal growth factor (EGF) receptor transinactivation and EGF receptor-independent stimulation of mitogen-activated protein kinase by bradykinin in A431 cells. Biochem J 347: 441–447
31. Mailman R B (2007) Ligand-selective signaling and high content for GPCR drugs. Trends in Pharmacol Sci 8:390–396
32. Samama P, Cotecchia S, Costa T, Lefkowitz RJ (1993) A mutation-induced activated state of the β2-adrenergic receptor. Extending the ternary complex model. J Biol Chem 268:4625–4636
33. Hermans E (2003) Biochemical and pharmacological control of the multiplicity of coupling at G-protein-coupled receptors. Pharmacol Ther 99:25–44
34. Lefkowitz RJ, Shenoy SK (2005) Transduction of receptor signaling by β-arrestins. Science 308:512–517
35. Azzi M, Charest PG, Angers S, Rousseau G, Kohout T, Bouvier M, Pineyro G (2003) β-arrestin-mediated activation of MAPK by inverse agonists reveals distinct active conformations for G protein-coupled receptors. Proc Natl Acad Sci USA 100:11406–11411
36. Sun Y, Huang J, Xiang Y, Bastepe M, Juppner H, Kobilka BK, Zhang JJ, Huang XY (2007) Dosage-dependent switch from G protein-coupled to G protein-independent signaling by a GPCR. EMBO J 26:53–64.

37. Nelson CP, Challiss RAJ (2007) "Phenotypic" pharmacology: the influence of cellular environment on G protein-coupled receptor antagonist and inverse agonist pharmacology. Biochem Pharmacol 73:737–751
38. Liapakis G, Chan WC, Papadokostaki M, Javitch JA (2004) Synergistic contributions of the functional groups of epinephrine to its affinity and efficacy at the β2 adrenergic receptor. Mol Pharmacol 65:1181–1190
39. Swaminath G, Deupi X, Lee TW, Zhu W, Thian FS, Kobilka TS, Kobilka B (2005) Probing the β2 adrenoceptor binding site with catechol reveals differences in binding and activation by agonists and partial agonists. J Biol Chem 280:22165–22171
40. Tran E, Fang Y (2008) Duplexed label-free G protein-coupled receptor assays for high throughput screening. J Biomol Screening 13:975–985

Adv Biochem Engin/Biotechnol (2010) 118: 165–188
DOI: 10.1007/10_2009_23

Published online: 21 January 2010

Cyanobacterial Bioreporters as Sensors of Nutrient Availability

George S. Bullerjahn, Ramakrishna Boyanapalli, Mark J. Rozmarynowycz, and R. Michael L. McKay

Abstract Due to their ubiquity in aquatic environments and their contribution to total biomass, especially in oligotrophic systems, cyanobacteria can be viewed as a proxy for primary productivity in both marine and fresh waters. In this chapter we describe the development and use of picocyanobacterial bioreporters to measure the bioavailability of nutrients that may constrain total photosynthesis in both lacustrine and marine systems. Issues pertaining to bioreporter construction, performance and field applications are discussed. Specifically, luminescent *Synechococcus* spp. and *Synechocystis* spp. bioreporters are described that allow the bioavailability of phosphorus, nitrogen and iron to be accurately measured in environmental samples.

Keywords Bioreporter • Cyanobacteria • Luciferase • Nutrients • *Synechococcus*

Contents

G.S. Bullerjahn, M.J. Rozmarynowycz, and R.M.L. McKay (✉)
Department of Biological Sciences, Bowling Green State University, Bowling Green, OH 43403, USA
e-mail: rmmckay@bgsu.edu

R. Boyanapalli
N30 Pharmaceuticals, Boulder, CO 80301, USA

1 Introduction

There is no photic environment on Earth in which the presence of cyanobacteria has not been documented [1]. From pole to pole, cyanobacteria can be major primary producers in both aquatic and terrestrial ecosystems. This is particularly true in oligotrophic marine and freshwater systems, where picocyanobacteria dominate the pelagic biomass (for reviews see [2, 3]). Indeed, it is estimated that, in the open sea, the picocyanobacteria *Prochlorococcus* spp. and *Synechococcus* spp. are responsible for up to 80% of the primary production [4–6]. Similar values have been documented for oligotrophic large lakes [7–9]. Given their ubiquity and position at the base of the food web, it comes as no surprise that cyanobacteria are viewed as excellent proxies for the productivity of the aquatic environment (e.g., [10, 11]). Such a fundamental position in the functioning of an ecosystem makes cyanobacteria ideal targets for identifying and monitoring factors that may constrain such productivity. As a consequence, genetically engineered cyanobacterial bioreporters have been developed for this purpose.

Specifically, a bioreporter is a cell designed to produce a quantifiable signal in response to a specific change in the environment of that cell [12, 13]. A regulated promoter element is fused to a gene whose product will yield a readout activity that can be monitored in real time. Typically, the readout is detected as luminescence from a luciferase activity [14, 15], or fluorescence of a GFP derivative [16]. Thus, environmental conditions can be examined from the perspective of a living cell, complementing parallel chemical and physical measurements of the environment in question. Overall, bioreporters provide information on the bioavailability or toxicity of a specific analyte to the cell, information that chemical measures of concentration and speciation cannot provide. In our laboratory and others, efforts have focused on engineering picocyanobacteria designed to report on the bioavailability of macro- and micronutrients, specifically phosphorus (P), combined nitrogen (N), and iron (Fe) [17–24]. Bioreporter construction and characterization has become increasingly straightforward, as there are a number of cyanobacterial strains amenable to genetic manipulation, and the wealth of genomic information has enabled the rapid identification and subcloning of specific target promoter sequences responsive to environmental stimuli. In this chapter, we will discuss the construction and use of cyanobacterial bioreporters, taking into consideration the reporter system, the strain to be employed, and the assay methods necessary to yield reliable data on the specific parameter to be measured.

2 Considerations for Bioreporter Construction

Characteristics of bioreporters vary depending on the experimental conditions and detection methods employed. The choice of reporter genes depends on factors such as expression efficiency, stability, background activity, detection method, and assay methodology [25]. Advantages of using bioreporter technology in environmental monitoring are sensitivity, selectivity, and stability, all of which provide ideal conditions to assay environmental samples. Another important advantage is their ability to provide physiological information pertaining to the whole cell that is not possible with methods involving cell extracts. Implementing the use of bioreporters can be a cost-effective, simple, and environmentally benign process, which facilitates on-site monitoring [26]. Limitations of bioreporters include concerns pertaining to their ecological relevance to a given system as well as situations where the analyzed compound may interact with extracellular material and alter the chemistry of an environmental sample [27]. Therefore, when choosing a bacterial reporter strain, it is vital to consider various environmental factors to design an informative reporter strain. In this section we will address the many variables to be encountered while developing a cyanobacterial bioreporter.

2.1 Reporter Genes

An initial consideration is the type of reporter gene to be employed. Whereas the majority of reporter fusions to date have been constructed with the *Vibrio harveyi luxAB* luciferase genes, by no means is that the only option. For example, GFP-based reporters have been developed in filamentous cyanobacteria to examine temporal gene regulation during heterocyst development (review, [28]), and platforms to yield GFP fusions for *Synechocystis* sp. strain PCC 6803 [16] and *Prochlorococcus* spp. [29] have been developed in the past decade. Additionally, eukaryotic luciferase genes (*luc, pxvGR, pxvRE*) have been employed to monitor circadian gene expression [15, 30]. Given the fact that bacterial and the varied eukaryotic luciferases have distinct chemistry, substrate requirements, and emission wavelengths, the characteristics of these enzymes allows for the development of multichannel bioreporters in which readouts from several promoters can be recorded from the same cell.

Each fusion gene and gene product has particular advantages, depending on the application. GFP and other fluorescent proteins require no substrate or redox chemistry, and thus the accumulation of the folded protein yields a direct readout signal. GFP and YFP derivatives typically emit at wavelengths (500–540 nm) that do not overlap with absorption spectra of photosynthetic pigments. However, in aquatic systems fluorescent cyanobacterial bioreporters are of limited utility when fluorescent organic matter is present that can yield a high signal-to-noise ratio (N.V. Ivanikova and G.S. Bullerjahn, unpublished data). This is of particular concern in fresh waters and coastal marine environments, where fluorescent CDOM (emission 400–550 nm) can comprise a significant fraction of total DOC [31–34]. By contrast, measurement of luminescence from luciferase fusions is not affected by interfering organic compounds, at least not directly (see [35] for discussion). Additionally, studies with *gfp* and *dsred* fusions in *Escherichia coli* have shown that bioreporter response is slower than luciferase constructs, possibly limited by the slow rate at which fluorescent proteins fold into their functional conformation [36]. In light of these issues, our use of *gfp* has been restricted to fusion constructs to a constitutive promoter (K.K. Brinkman and G.S. Bullerjahn, unpublished). In concert with a regulated luciferase reporter fusion, the population size of the reporter strain in an environmental sample can be monitored by GFP fluorescence in a flow cytometer or by epifluorescence microscopy.

Luciferase fusions, both *luxAB* and *luc*, involve the expression of a monooxygenase that requires an organic substrate and molecular oxygen. Even though the chemistry of the two classes of enzymes are vastly different, the nature of luciferase fusions requires that the bioreporter cells are metabolically active and in an aerobic environment. Availability of the appropriate substrate is also an important consideration. In the case of bacterial luciferase, the *luxCDE* genes direct the synthesis and regeneration of the acyl aldehyde substrate. In the pioneering work by Susan Golden's lab, bacterial luciferase fusions were generated to explore *Synechococcus elongatus* PCC 7942 circadian gene expression. In such strains, the *Vibrio* sp. *luxCDE* genes are under constitutive control, allowing endogenous substrate for long-term real-time monitoring of luminescence [37]. However, such constructs often do not produce saturating concentrations of substrate, requiring addition of exogenous decyl aldehyde to maximize luminescence [18, 38]. To date, cyanobacterial *luc, pxvGR* and *pxvRE* fusions bearing the eukaryotic luciferase genes have relied on exogenous luciferin added to the assay medium [15, 30, 39] as well as an energy requirement in the form of ATP. Bacterial luciferase emits at 490 nm [40], and eukaryotic luciferases PxvGR, Luc, and PxvRE emit in the green (549 nm, 560 nm) and red (622 nm), respectively, allowing spectral discrimination among different promoter fusions in the same cell [15, 30].

Whereas *luc* and other eukaryotic luciferases are important tools in the construction of multichannel cyanobacterial bioreporters, our lab has traditionally relied on the construction of *luxAB* promoter fusions. Such strains yield a robust luminescent signal detectable from bioreporter densities mimicking that of pelagic picocyanobacterial populations (10^5 mL^{-1} or ca. 1 μg chl *a* L^{-1}). Below we describe the construction and use of freshwater and marine bioreporters to assess aquatic nutrient bioavailability.

2.2 Construction Strategies

Depending on the regulation of the promoter to be employed in bioreporter construction, luciferase-dependent reporters can be designed to respond specifically to an analyte yielding either an increase in luminescence (Class I, or "Lights on") or a decrease in light emission (Class II, or "Lights off") [13, 26]. In our laboratory, we have constructed both types and successfully used them to assess Fe (Class II; [17–19]) and nitrate bioavailability (Class I; [22]). In this section we discuss strategies for bioreporter construction, taking into consideration the host strain and gene transfer protocols.

2.2.1 Genetic Transformation Strategies: Prototype Reporter Strains

The first cyanobacterial reporter strains constructed used the freshwater model strains *Synechococcus elongatus* PCC 7942 [18–21] and *Synechococystis* sp. PCC 6803 [22] and the halotolerant *Synechococcus* sp. PCC 7002 [17]. To date, bioreporter strains we have developed and used with success include Fe-dependent Class II *Synechococcus* spp. PCC 7942 and PCC 7002 bioreporters that utilize the Fur-regulated *isiAB* promoter [17, 18], and the *irpA* promoter, part of the Fe-responsive IdiB regulon [41–43]. Also developed has been a *Synechocystis* sp. PCC 6803 Class I reporter employing the NtcA/NtcB-regulated *nirA* (nitrite reductase) promoter [22]. Such strains have been employed to assess Fe and nitrate bioavailability in environmental samples (see Sect. 3). Since these strains can be manipulated by efficient natural genetic transformation, promoter fusion plasmids can be constructed and introduced into host strains as naked DNA. Furthermore, the complete genome sequences available from these strains allow the amplification of any promoter for fusion construction. Given the ease of this approach, and an existing genetic toolbox of *luxAB* promoter probe plasmids [14–16], a suite of *Synechococcus* PCC 7942 and *Synechocystis* sp. PCC 6803 bioreporters were rapidly developed and characterized. Modification of the *S. elongatus* promoter fusion plasmid pAM1414 [14, 15] yielded a suite of similar vectors for *Synechococcus* sp. PCC 7002 [17]. Despite the fact that these strains are not particularly ecologically relevant with respect to most oligotrophic aquatic environments, these prototype strains afforded an opportunity to demonstrate the feasibility in using reporters in both freshwater [9, 18, 35, 38, 41, 44–46] and marine samples [17].

In general, the construction strategy involves the use of a plasmid vector incapable of replication in the cyanobacterial recipient strain. Chromosomal sequences engineered into the plasmid that flank a promoter::*luxAB* fusion and an antibiotic resistance marker allow the selection of drug-resistant survivors that have acquired a stable copy of the fusion construct through homologous recombination [14–17, 37]. Often, the site for integration into the chromosome is a so-called neutral site (NS), chosen because integration yields no observable phenotype [15, 16]. Indeed, combining two or more NS integrative vectors allows the construction of strains that express a promoter fused to *luxAB* at one site (NS1) and *luxCDE* under constitutive expression at another (NS2). Alternatively, multiple luciferases can be expressed by combining several NS insertions in the same strain [15]. By contrast, recombination sites can be designed that yield a useful discernable phenotype. For example, the *Synechococcus* sp. PCC 7002 promoter probe plasmids recombine into the *desB* gene, encoding a fatty acid desaturase [17, 47]. The resulting construct is incapable of growth at temperatures below 15°C, thereby reducing the likelihood that the strain will grow if inadvertently released into the field. Similarly, the strategy of constructing a *Synechococcus* sp. WH8102 reporter (described below) yields a phenotype useful in assessing the stability of the genetic construct.

2.2.2 Conjugative Strategies for Marine *Synechococcus* spp.

Our long-term goal has been focused on the development of a suite of picocyanobacterial reporters in strains that are ecologically relevant to the pelagic open sea. Complicating this aim is the lack of a genetic transformation system for marine *Synechococcus* spp. or *Prochlorococcus* spp. As a result, plasmid constructs are delivered into *Synechococcus* spp. by conjugation from an *E. coli* donor (see below). Additionally, considerations must be made regarding the type of strain to employ as a bioreporter host strain. Indeed, the availability of several picocyanobacterial genome sequences yields a potentially difficult choice. Ideally, the best candidate would be a genetically manipulable strain whose genome has been sequenced, and for which nutrient assimilation pathways have been analyzed in silico [48, 49]. Meeting all these criteria is the strain *Synechococcus* sp. WH8102, a motile, pelagic marine generalist [48] and a member of the picocyanobacterial clade that includes both marine *Synechococcus* spp. and *Prochlorococcus* spp. [50]. Thus, we are developing bioreporter systems in this strain by mobilizing *luxAB* promoter fusions by conjugation.

As mentioned above, marine picocyanobacteria cannot be manipulated by genetic transformation. Instead, foreign DNA is introduced by conjugation from *Escherichia coli* donors bearing derivatives of broad host range plasmids capable of efficient transmission among divergent bacterial taxa (e.g., [51, 52]). The *E. coli* donor strain contains a plasmid (e.g., pRK24) bearing the broad host range transfer functions required for conjugation. Indeed, efficient conjugational transfer of DNA has been documented for *Synechococcus* sp. WH7803, WH8102, and WH8103 (3–6 $\times$ 10^{-3} per recipient; [51]) and *Prochlorococcus* sp. MIT9313 [29]. For conjugation into these strains, broad host range plasmids derived from IncP (RP4 or RK2) and IncQ (RSF1010) incompatibility groups have been employed, and thus these plasmids can be used as templates for the development of transmissible vectors bearing a *lux* promoter fusion capable of recombination into the chromosome. Alternatively, the promoter fusion can be delivered on a plasmid that can replicate in the recipient *Synechococcus* sp. strain at low copy number. Specifically, IncQ plasmid pRL153 has been used in matings with *Synechococcus* spp. [51] and *Prochlorococcus* sp. MIT9313 [29] to yield an extrachromosomal promoter fusion plasmid. The availability of this plasmid system allows a rapid means for delivering and testing prototype promoter fusions into *Synechococcus* sp. WH8102 with high efficiency (Fig.1).

To yield a suicide plasmid capable of recombination with the chromosome, an RP4 or RK2 plasmid derivative maintained in the *E. coli* donor strain provides broad host range transfer functions allowing the delivery of a second plasmid into the cyanobacterial recipient. For our purposes, *E. coli* BW20767 bearing plasmids pRK24 and pRL528 [51, 53] is suitable to provide the transfer functions. To achieve this, a number of sequences must be present on the suicide plasmid. For conjugative transfer to WH8102, an appropriate *oriT* (RP4/R6K) compatible with the RP4 transfer functions provided by the *E. coli* donor strain [52] is employed on the delivery suicide plasmid. The plasmid also contains a convenient cloning site

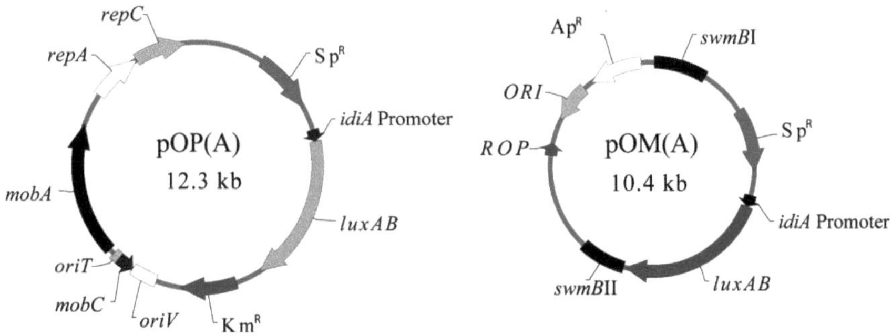

Fig. 1 Map of plasmid pOP (*left*) and pOM (*right*). Both plasmids are drawn to include the Fe-responsive *Synechococcus* sp. WH8102 *idiA* promoter, yielding pOP(A) and pOM(A). pOP is an extrachromosomal element replicating at low copy number. pOM is a suicide plasmid designed to promote the integration of the promoter fusion into the *Synechococcus* sp. WH8102 *swmB* gene by homologous recombination

adjacent to the promoterless *luxAB* genes and an antibiotic resistance marker, which are in turn flanked by chromosomal sequences allowing site-specific insertion into the chromosome by homologous recombination. Promoter fusion constructs are identified as spectinomycin resistant survivors arising following patch conjugation on SN plate media and subsequent selection on antibiotic-containing pour plates [53].

Last, consideration must be given to the recombination site into the WH8102 genome. Ideally, the promoter-probe platform insertion should yield a neutral phenotype with respect to the physiological processes assayed by the bioreporter. As a first step, we have used internal sequences of the *swmB* gene (ca. 800 bp flanking DNA) as a site for recombination (Fig. 1). *swmB* is a large gene (32 kb) essential for swimming motility in WH8102, a trait not widely distributed among the picocyanobacterial clade [53]. Due to the low GC content of *swmB* compared to the rest of the genome, it has been proposed that the gene was acquired by this strain via horizontal gene transfer [53]. We employ *swmB* as a recombination target because interruption of the gene does not affect the strain's capability to function as a bioreporter, yet the genetic stability of the chromosomal insertion can be assessed by examining the behavior of the construct in pour plates. A stable site-specific insertion yields small, nonmotile colonies, phenotypically distinct from the wild type.

Expanding bioreporter construction to other *Synechococcus* strains inhabiting distinctly different marine environments is another long-term goal. For example, *Synechococcus* CC9311 is well adapted to a coastal environment by bearing a large complement of genes encoding functions for metal ion utilization and storage [54]. Conversely, fewer genes comprise the P regulon encoding enzymes involved in phosphorus scavenging, reflecting the generally higher P bioavailability in coastal regions. Since CC9311 is also amenable to genetic manipulation by conjugation, a suite of reporter constructs in this strain would provide additional insights into how coastal members of the picocyanobacterial clade sense nutrient fluxes compared to pelagic forms. This additional approach would also help determine the long-term validity of PCC 7002 strains that we have successfully manipulated as bioreporters for coastal environments.

3 Cyanobacterial Bioreporters as Environmental Sensors

3.1 Bioreporters of Trace Metal Availability: Fe

The role of Fe in physiological processes such as photosynthesis, respiration, and nitrogen assimilation makes it one of the most important nutritive factors for phytoplankton growth. Despite its high abundance in the Earth's crust, low Fe availability has been shown to limit phytoplankton growth in diverse marine environments [55, 56]. This apparent contradiction is attributed to features of Fe biogeochemistry that lead to precipitation or complexation of Fe species in oxic waters. These considerations, combined with regionally low aeolian input of Fe to many open ocean environments [57], result in Fe deficiency to the endemic phytoplankton.

Efforts focused on studying Fe biogeochemistry over the past several decades have been biased toward global surveys, providing reliable measures of total dissolved (DFe) and particulate Fe (PFe). Data collected show that DFe is present often in subnanomolar concentrations in open ocean surface waters. Although absolute measures of DFe have been made in these regions, the proportion of the total Fe pool that is available to the phototrophic plankton is poorly known. Further, whereas chemical measures of total DFe can provide an estimate of the potential for Fe limitation, DFe need not be synonymous with bioavailable Fe. Some forms of PFe appear to be bioavailable [35, 44] whereas some Fe associated with the operationally-defined dissolved phase ($<0.45\ \mu m$) is not immediately available for uptake [58]. The complex chemical speciation of Fe in aquatic systems and the uncertainties associated with biological assimilation of Fe species make it difficult to ascertain the fractions of chemically detectable Fe that are readily available to phytoplankton.

Numerous approaches have been used to measure Fe availability in aquatic systems. Enrichment bioassays, though offering direct experimental evidence for growth limitation, do not completely mimic the undisturbed natural environment: grazing is disrupted, physical mixing is decreased, and the phytoplankton are isolated at a fixed optical depth [59]. To overcome these concerns, several biochemical and molecular approaches have been suggested such as monitoring the expression of Fe-responsive genes in environmental samples [60, 61], measuring the ratios of the redox catalysts ferredoxin:flavodoxin [62, 63] or other suitable Fe-responsive proteins such as IdiA [64], and assaying variable chlorophyll fluorescence [65]. Implementing a living system such as a bioreporter organism can help us gain a better understanding of the availability of Fe from the perspective of a living cell.

A prototype Fe-responsive bioreporter was developed from the unicellular freshwater cyanobacterium *S. elongatus* PCC 7942 in a construct bearing a genetic fusion of the Fe-responsive *isiAB* promoter to the *Vibrio harveyi luxAB* genes encoding bacterial luciferase [18]. P*isiAB* is regulated in part by the Fe-dependent repressor Fur, yielding a Class II ("Lights off") reporter strain [13]. Characterization of this strain, KAS101, demonstrated the luminescent response to be a function of the free ferric ion concentration in metal-buffered media [38, 41, 66]. Further standardization of assay conditions has allowed the use of this strain in documenting Fe bioavailability in the Laurentian Great Lakes [9, 35, 38, 46]. While the use of Fe-responsive bioreporters has offered some evidence of transient Fe deficiency in the Great Lakes (e.g., [9, 62]), it is not widespread [35, 46]. A new approach whereby the bioreporter is contained in a porous underwater chamber (PUC) during incubation with unfiltered water [44] promises to offer insight into the biological cycling of Fe by regenerative processes such as grazing and viral lysis which may provide 30–80% of the algal Fe demand in surface seawater [62].

In an effort parallel to ours, Kunert et al. [16] developed *luxAB* and *gfp* constructs regulated by the promoter for *isiAB* in *Synechocystis* sp. PCC 6803. As expected, bioreporter luminescence and fluorescence increased in response to Fe deficiency. Both bioreporters also reacted to high salt stress, a response likely related to oxidative stress which appears to be a common control on *isiAB* transcriptional regulation in cyanobacteria [67].

The repertoire of cyanobacterial Fe-responsive bioreporters was recently expanded by development of a luminescent whole-cell cyanobacterial bioreporter, strain BMB04 (deposited as *Synechococcus* sp. CCMP 2669), constructed using the euryhaline *Synechococcus* sp. PCC 7002 for measuring Fe availability in diverse marine environments [17]. A dose-response curve was generated relating bioreporter luminescence to the free ferric ion content of defined growth medium between pFe 19.4 and 22.4 (see Fig. 1 of [17]). Through this range of variable [Fe^{3+}], discernible changes were measured in the luminescent response of cells with luminescence being >2× higher associated with cells growing under low Fe (pFe 22.4) compared to Fe sufficient conditions (pFe 19.4). Luminescence plotted as a function of pFe could be described according to a three-parameter sigmoidal curve and, as such, provided a similar response to that which we have characterized for a suite of freshwater Fe bioreporters (e.g., [38]).

3.1.1 Case Study: Assessing Fe Bioavailability in the Central North Pacific Gyre

The halotolerant Fe bioreporter *Synechococcus* sp. CCMP 2669 was successfully used to determine Fe bioavailability in samples from the Baltic Sea and during the SERIES mesoscale Fe fertilization conducted in the vicinity of Ocean Station Papa in the subarctic Pacific Ocean [17]. This reporter strain has also been used to assess Fe availability along a transect (27–28°N) in the central North Pacific (CNP) gyre. The CNP gyre is the largest circulation feature on Earth with a surface area of ~2×10^7 km^2 extending across the subtropical North Pacific Ocean. A classic oligotrophic ocean gyre, surface waters of the CNP are permanently stratified with mixed layer temperatures >24°C. Mixed layer nitrate concentrations along our transect were ~2 nmol L^{-1}, and surface ammonium concentrations were <30 nmol L^{-1} [68, 69]. Likewise, dissolved inorganic phosphate is depleted with concentrations ranging between 21 and 150 nmol L^{-1} reported from the mixed layer at the time-series oceanographic station ALOHA (22.75°N, 158°W) [70, 71]. DFe concentrations in surface waters of the CNP gyre are also very low, in the range of ~50–80 pmol kg^{-1} [72, 73], although inputs vary both temporally and spatially. DFe in surface waters can originate from atmospheric dust, diffusive flux from deep waters, and biological regeneration. Typical of most oligotrophic systems, the surface waters of the CNP are dominated by picocyanobacteria, such as *Prochlorococcus* and *Synechococcus* spp. [74, 75]. These planktonic communities are able to compete effectively for predominantly regenerated nutrients, in part due to their large surface area to volume ratios. Using the bioreporter to assess Fe availability in filtered water samples collected from each station, we observed meaningful differences in Fe availability along the E→W transect (Fig. 2). Specifically, as

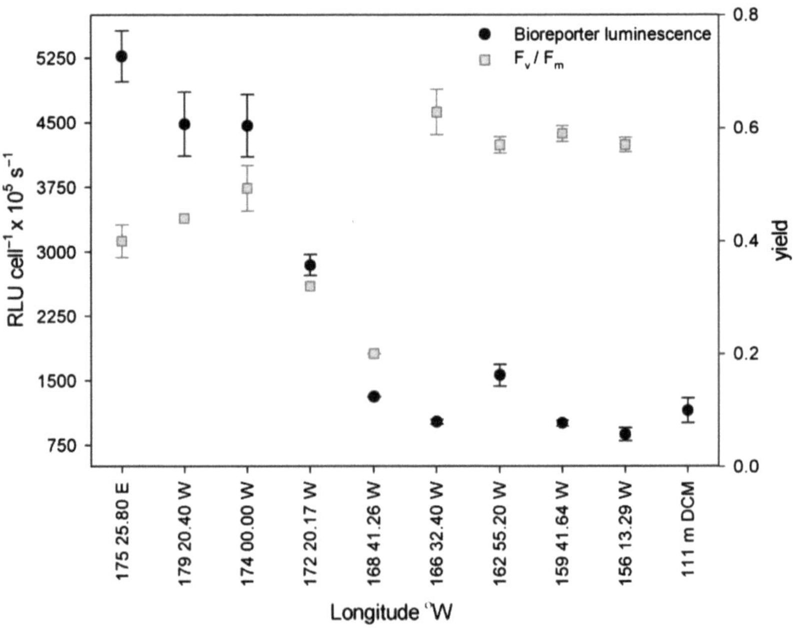

Fig. 2 For bioreporter analysis, seawater was collected in acid-cleaned Teflon bottles from 1–2 m below the surface by divers operating from an Avon inflatable boat at a distance >500 m from the research vessel. The water was filtered (<0.2 μm) in a laminar flow bench on the research vessel using a Teflon filtration assembly. All manipulations were conducted using trace metal clean techniques. The Fe-dependent luminescent response of the *Synechococcus* sp. CCMP 2669 bioreporter was assessed by incubation in samples of filtered water as described previously [17]. Buoyant aggregates (mats) of the diatom *Rhizosolenia* spp. were collected at each site by divers with analysis of quantum yield (F_v/F_m) made by Pulse Amplitude-Modulated fluorescence as described elsewhere [68]

we proceeded west along the transect, bioreporter luminescence increased >4× from station 5 (168.7°W) to station 9 (175.4°E) (two-tailed *t* test; $P < 0.0005$). This is interpreted as decreasing Fe availability at locations west of ~168°W.

Along this same transect, we measured variable chl fluorescence (F_v/F_m) for the buoyant diatoms *Rhizosolenia* spp. [68] and *Ethmodiscus* spp. [69]. To the east of 168°W, *Rhizosolenia* F_v/F_m was near maximal at ca. 0.6 [68] whereas the bioreporter response was low, indicating higher Fe availability compared to the waters west of 168°W (Fig. 2). West of this longitude, F_v/F_m decreased to ca. 0.4, indicative of cells that were physiologically stressed. Likewise, the response of the cyanobacterial bioreporter was similarly indicative of physiological stress by, in this case, low Fe availability. Limitation by Fe is often reflected in depressed F_v/F_m measured for phytoplankton [76]. Thus, the measured pattern of *Rhizosolenia* F_v/F_m was consistent with Fe deficiency.

While no DFe data are available from this study, the results of past studies conducted in the CNP gyre offer a general interpretation of our results. In the CNP, the distribution of Fe appears as a macronutrient-type profile [72] where concentrations of DFe are elevated in deep waters compared to levels measured near the surface. Surface mixed layer concentrations of 0.02–0.38 nmol kg^{-1} have been measured at 28°N and 155°W, about 200 km north of Hawaii (Vertex IV station) [77]. Likewise, during the 2002 Intergovernmental Oceanographic Commission (IOC) cruise, surface mixed layer DFe concentrations ranged between 0.04 and >1 nmol kg^{-1} (median: 0.13 nmol kg^{-1}; $n = 21$) along a transect that essentially paralleled our study transect [78]. Further, there was a distinct trend showing an increase in DFe along the eastern reaches of this transect with a median value of 0.45 nmol kg^{-1} ($n = 5$) measured for DFe between 155 and 160°W [78]. Thus, several lines of evidence suggest a gradient of decreasing Fe availability towards the west including DFe concentration [78], decreasing *Rhizosolenia* F_v/F_m [68], and our studies using the Fe-responsive marine cyanobacterial bioreporter.

We also compared Fe availability at depth with a surface mixed layer sample from a single station (Station 8: 174°W). As stated earlier, vertical profiles of DFe generally exhibit a classic nutrient-type profile [72] consistent with remineralization of Fe from detritus at depth. Consistent with this, bioavailable Fe as determined using the cyanobacterial bioreporter was higher at depth (111 m) compared to surface water. This was reflected by a nearly 4× lower bioluminescent signal elicited from the deep water sample compared to a surface sample (two-tailed *t*-test; $P < 0.001$).

3.2 Bioreporters of P Availability

Phosphorus has been described as the "staff of life – the most essential of nutrients" owing to its importance to biota and the low ambient concentrations at which it is present in many surface waters [79]. Phosphorus is an essential component of DNA, ATP, and phospholipids and accounts for about 2–4% of the dry weight of most cells [79]. Yet in many aquatic environments, inorganic phosphate (DIP) is near detection limits using standard analytical methods.

3.2.1 Issues and Challenges in Assessing Bioavailable P

Whereas the limitation of primary production by P availability is a central tenet of modern day limnology, it has only recently been afforded more widespread attention among oceanographers (see review by [80]). This stems, in part, from a long-held belief that P limits production only over prolonged time scales. Over shorter periods, N is thought to constrain phytoplankton growth in the open ocean. This is supported by constraints imposed on phytoplankton nutrient requirements by the canonical Redfield ratio and consideration of ratios of inorganic N (DIN): DIP that show exhaustion of N before P [81]. Neglected in such a consideration, however, is the potential availability of organic nutrients to phytoplankton. Indeed, when total dissolved P (TDP) is plotted against total dissolved N (TDN) and the trend is extrapolated to nutrient exhaustion, TDP is depleted before TDN [81]. This is important when considering that dissolved organic phosphate (DOP) represents the dominant fraction of TDP in the upper oligotrophic ocean [82, 83].

Contributing further to our lack of understanding of P bioavailability are the somewhat arbitrary methods used to measure P (see review by [80]). Most commonly used is the acid molybdate method capable of measuring P to 30 nM. A modification of this approach, the magnesium-induced coprecipitation (MAGIC) procedure, is the most sensitive chemical technique, detecting P to <5 nM [71]. Although long assumed that the acid molybdate method measures orthophosphate, it is now recognized that formation of the phosphomolybdate complex during sample acidification is accompanied by hydrolysis of some fraction of the DOP pool. Thus, it is more appropriate to refer to the fraction measured using this technique as soluble reactive phosphorus (SRP). To measure TDP, samples are exposed to high temperature and/or pressure in the presence of a strong oxidizing agent prior to phosphomolybdate complex formation. DOP can then be inferred as the difference between TDP and SRP. Still, interpretation is not straightforward since the presumed DOP fraction may contain some nonreactive inorganic species such as polyphosphates. Alternatively, some components of the DOP fraction may be refractory and are not readily converted to TDP. For example, naturally-occurring phosphonates comprise up to 25% of the high molecular weight DOP pool in the open ocean [84], and are generally thought to represent a P source more refractory to assimilation than organic monophosphate esters [80].

Overall, it is clear that current techniques to measure P provide little information on speciation. Besides the conventional colorimetric methods, several alternative strategies of P determination based on ion-selective electrodes or enzyme sensors (see review by [85]) have been developed in recent decades. Complementary to these approaches are bioassays to assess the utilization of known or unknown dissolved P species (see review by [86]). One of the widely used methodologies is the (Provisional) Algal Assay Procedure (as described in [86]) whereby a P deficient monoalgal culture is inoculated in a sample and algal yield recorded after several weeks. The yield is converted to P equivalents by calibration with parallel trials run with DIP although this approach suffers from many of the complications associated with experiments requiring prolonged incubations.

Many bacteria and phototrophic plankton can exploit forms of DOP, mainly through enzymatic reactions at the cell surface. Fluorometric ectoenzyme assays offer modest resolution between DOP species and afford a quantitative method to measure P availability. Two major enzyme classes appear to be most important in this capacity: alkaline phosphatase, an inducible monophosphate esterase having broad substrate specificity [87] and 5′-nucleotidase, an enzyme capable of hydrolyzing the carbon moiety of nucleotides, releasing DIP in the process [88]. A modification of the bulk alkaline phosphatase approach known as enzyme-labeled fluorescence (ELF) affords cell-specific detection of P stress in mixed populations [89, 90].

Additional biochemical and molecular approaches that have been employed include monitoring the expression of the high affinity periplasmic P-transporter PstS in P-deficient marine cyanobacteria both by immunoblotting [91–93] and by reverse transcriptase (RT)-PCR using *pstS*-specific primers [94]. Recently, the expression of genes involved in phosphonate transport has been demonstrated in laboratory and field populations of the marine diazotroph *Trichodesmium* [95] and among picocyanobacteria [96], thus countering the prevailing view that phosphonates are refractory sources of P.

3.2.2 P-Responsive Bioreporters

The use of compound-specific P bioreporters may offer a unique approach toward not only assessing the bioavailability of various P chemical species but also afford the opportunity to characterize readily the components of the P pool, especially DOP species at a given site. At present, investigators are constrained in their ability to characterize the DOP pool by the requirement to concentrate large volumes of water using techniques such as tangential flow ultrafiltration or lyophilization followed by ^{31}P nuclear magnetic resonance (e.g., [84]). By exploiting specific genetic mechanisms used by cyanobacteria to target individual components of DOP, bioreporters offer a means by which to characterize the DOP pool using small volumes of water (as little as 100–200 μL if adapted to a microtiter plate assay). Moreover, since bioreporter response can be calibrated using model DOP compounds, the concentration of specific DOP classes in natural waters can be determined.

Applicable to monitoring PO_4-Pi deficiency in aquatic environments, a luminescent cyanobacterial P bioreporter was designed by Gillor et al. [20]. This bioreporter, appropriate for use in freshwater environments, was constructed using a P*phoA::luxAB* construct integrated into the cyanobacterium *Synechococcus elongatus* PCC 7942 and features a detection range of 0.3–8.0 μmol L^{-1} PO_4-Pi using a sample incubation time of 8 h. The bioreporter was designed using the promoter for *phoA*, a P-responsive gene that encodes alkaline phosphatase. The resulting strain responds by dose-dependent light emission to a wide range of PO_4-Pi concentrations. This freshwater P bioreporter has also been adapted for use as an immobilized sensor in microtiter plates called the "Cyanosensor" [24].

Effort is currently being allocated to developing compound-specific P bioreporters in both *Synechococcus* sp. PCC 7002 and WH8102 able to resolve DIP and DOP, including nucleotides and phosphonates as P sources. The suite of P bioreporters developed should have the ability to discriminate between use of DIP and DOP and should afford resolution of major classes of DOP utilized by aquatic photoautotrophs.

3.3 Bioreporters of N Availability

As demonstrated for Fe and P, the bioreporter approach can be viewed as an alternative method to measure nitrogen uptake in aquatic systems. With this rationale, a *Synechocystis* sp. strain PCC 6803 bioluminescent reporter was developed capable of assessing nitrate assimilatory capacity in freshwaters [22]. The bioreporter, designated AND100, is based on the nitrate/nitrite-activated *nirA* promoter (regulating expression of genes encoding nitrite reductase) which is under positive control by two transcription factors, NtcA and NtcB, that together yield elevated transcription when bioavailable nitrate or nitrite is present in the medium [97, 98]. Characterization of this strain demonstrated a dose-dependent increase in bioluminescence coincident with increased nitrate added to the growth medium to 100 μM. Additionally, bioluminescence in response to nitrate addition was light dependent up to 50 μmol quanta m^{-2} s^{-1}. Because nitrate concentrations in most freshwater systems exceed those of nitrite by more than an order of magnitude (e.g., [99]), the bioreporter can be viewed primarily as a sensor for nitrate bioavailability.

The AND100 strain differs fundamentally from two cyanobacterial nitrogen bioreporters previously described [21, 23]. These bioreporters are *luxAB* fusions employing the *Synechocystis* sp. PCC 6803 *nblA* [23] and *S. elongatus* sp. PCC 7942 *glnA* promoters [21], controlling the genes encoding a phycobilisome degradation regulator and glutamine synthetase, respectively. Whereas the dynamic ranges of these strains were similar to AND100, the luminescent response was induced upon nitrogen deficiency, not during nitrogen use as shown for AND100. Response time of the *S. elongatus* sp. P*glnA* and *Synechocystis* sp. P*nblA* bioreporters was also considerably slower, with each reporter strain yielding dose-dependent luminescence on the order of 15–25 h [21, 23] compared to 4 h for AND100. Additionally, the *glnA* strain yielded dose-dependent responses to a wide variety of N species such as nitrate, ammonium, urea, and glutamine [21], and *nblA* expression was responsive to nitrate and ammonium [23]. Such broader spectrum responses may be useful properties for the measurement of total nitrogen bioavailability, and indeed, the P*glnA* reporter has been used to document decreasing nitrogen bioavailability along a west-to-east transect in Lake Erie [100].

By contrast, the nitrate/nitrite specificity of the AND100 bioreporter provides a means for discriminating between specific nitrogen species. In this context, the AND100 bioreporter likely could provide further insights by focusing on the potential for nitrate use in the Great Lakes. Nitrate has been accumulating in the Laurentian Great Lakes for decades. Lake Superior represents an end member in this regard with a century-long sixfold increase resulting in present-day nitrate accumulation approaching 30 μmol L^{-1} [101]. Whereas seasonal drawdown of nitrate is demonstrated, it is modest (2–4 μmol L^{-1}) compared to the lower lakes in the Great Lakes system [101]. To gain insight into the factors that constrain nitrate consumption in Lake Superior, the positive induction of the AND100 luminescent response was used to measure the onset of nitrate use as light levels were manipulated and nutrients (P, Fe) were amended to samples collected from the lake [8, 45]. These data suggest that P- and Fe-limited cyanobacteria are deficient in their ability to assimilate nitrate in Lake Superior (Fig. 3a). Furthermore, during spring, light fluxes are sufficiently low to prevent maximal nitrate utilization, even in the absence of nutrient limitation (Fig. 3b). By comparison, the properties of the *Synechococcus* sp. P*glnA* bioreporter, whose bioluminescence is under repression by elevated nitrogen, would not be suitable for such an experiment. Overall, the AND100 strain affords a direct method for determining the role of both chemical and physical factors in regulating nitrate uptake by photoautotrophs.

Whereas the aforementioned strains were developed expressly as environmental sensors, manipulation of cyanobacteria to produce luminescent transcriptional reporters to elucidate components of regulatory networks is a strategy that has been adopted previously (reviewed by [102]). Of relevance to N assimilation, Wolk et al. [103] used a derivative of transposon Tn5 with the filamentous, diazotroph *Anabaena* sp. PCC 7120 to generate transcriptional fusions to promoterless bacterial luciferase genes. Using this approach, they identified genes that responded to removal of fixed nitrogen by monitoring the luminescence of colonies from transposon-generated libraries. Likewise, luminescent transcriptional reporters have been developed to study NtcB-regulated expression [104] of genes involved in N assimilation in *S. elongatus* sp. PCC 7942. Application of these strains as environmental sensors is a reasonable extension.

4 Algal Bioreporters

We recognize that the nutrient stress responses of the strains we have used may not adequately reflect the response of phytoplankton taxa in the diverse aquatic environments in which nutrient deficiency has been documented. Behind our rationale for developing cyanobacterial bioreporters is that the prospective host strains described have been the subject of genome sequencing and annotation efforts, they are well characterized with respect to nutrient stress responses, and they are amenable to genetic manipulation.

Yet, by relying solely on a prokaryotic bioreporter, we cannot comment with certainty on nutrient availability to eukaryotic phototrophs (microalgae). Considering the dichotomy in nutrient acquisition strategies between prokaryotes and

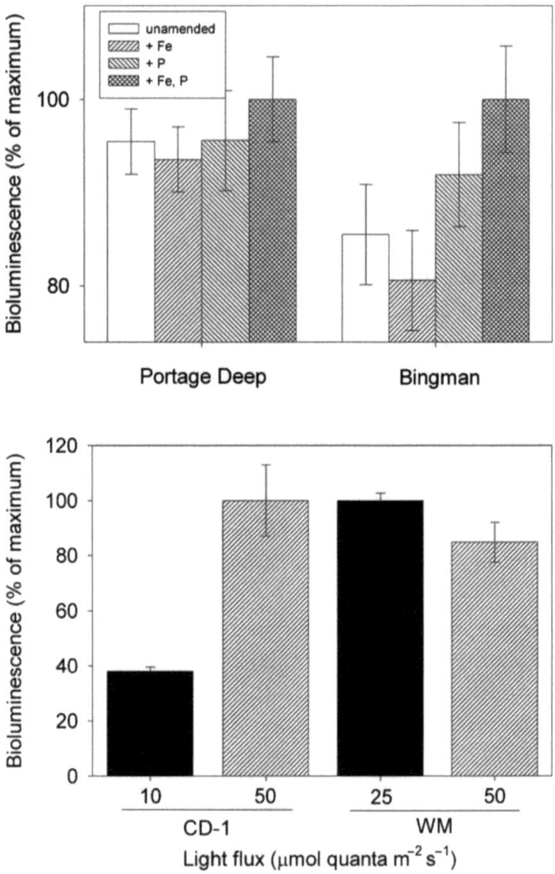

Fig. 3 Use of the N-responsive bioreporter AND100 to gain insight into abiotic factors that constrain nitrate consumption in Lake Superior. (**a**) Comparison of bioreporter response between water sampled in September 2004 from the P, and Fe- sufficient Portage Lake (Station Portage Deep) and in the open waters of Keewenaw Bay (Station Bingman). Nutrients were added to 10 nmol kg^{-1} Fe and 8 μmol L^{-1} P where indicated. (**b**) Effect of light treatment on bioreporter response. Water samples collected during May, 2004 from Stations CD-1 and WM were seeded with the nitrate bioreporter and luminescence was measured following 7 h incubation at the calculated light fluxes representing integrated spring mixing and summer stratification. Figures republished with permission of Blackwell Publishing Limited

eukaryotes for elements such as Fe [58], development of suitable eukaryotic nutrient-responsive bioreporters is warranted. However, in doing so, one is confronted with present limitations on the genetic engineering of microalgae [105, 106] where success has been largely limited to a few model species such as *Chlamydomonas reinhartii* where the development of nuclear and organellar transformation systems are highly advanced [107]. Whereas *Chlamydomonas* nutrient-responsive bioreporters have not been developed *per se*, several transformants could be applied as such. Recently, Ferrante et al. [108] describe development of a *CYC6:cRLuc* construct inducible by copper (Cu) depletion and proposed for use in the development of efficient, inducible expression systems facilitating the use of these algae as "green cell-factories" for the production of value-added metabolites and proteins [109, 110]. Under Cu-deficiency, the electron transport protein cytochrome c_6, encoded by *cyc6*, replaces the Cu-containing redox catalyst plastocyanin in green algae and cyanobacteria [111]. In the construct described in Ferrante et al. [108], Cu-deficit results in expression of a synthetic luciferase designed from the sea pansy *Renilla reniformis* modified in compliance with the average codon usage of nuclear genes from *C. reinhardtii* [112].

Recent success at stable nuclear transformation of some marine diatoms (as reviewed by [113]) combined with ongoing genome sequencing and annotation efforts for several diatom species [114, 115] has cleared some of the obstacles to developing a diatom bioreporter organism, thus offering the potential of complementary bioreporter systems representative of the diversity of nutrient acquisition strategies exhibited by phytoplankton. Falciatore et al. [116] describe a transgenic diatom, *Phaeodactylum tricornutum*, designed to study the perception of environmental signals. The expression of the reporter protein, aequorin, responded to physicochemical changes in the extracellular environment yielding fluctuations in cytoplasmic calcium concentrations. Of the various nutrient stresses assessed, the reporter gene was expressed only in response to Fe deficiency, yet this particular reporter would not be suitable for environmental monitoring since the threshold of reporter gene expression was elevated substantially (5–10 μmol kg^{-1}) compared to the low (sub-) nanomolar Fe conditions reported in most marine environments.

Recently, inducible reporter gene expression under the control of promoter and terminator elements of the gene encoding nitrate reductase (*NR*) was reported in transgenic diatoms *Cylindrotheca fusiformis* [117] and *Thalassiosira pseudonana* [118] as well as in the green alga *Dunaliella salina* [119]. In each case, (enhanced) green fluorescent protein accumulated in cells grown with nitrate, but not when ammonium was included in the medium, thereby demonstrating appropriate N-responsive control by the *NR* promoter over the expression of the reporter gene. While the primary motivation behind these efforts was to develop their use as "green cell-factories" as well as for the production of inorganic nanomaterials [120, 121], we recognize their parallel use as N-responsive bioreporters, complementary to the cyanobacterial reporters described earlier.

5 Concluding Remarks

Work over the past decade has shown that cyanobacteria can be engineered to yield reliable data on the bioavailability of nutrients in environmental samples. Genetic toolboxes have been developed to enable the construction of transcriptional fusions in a wide variety of cyanobacterial genera endemic to freshwater and marine habitats.

Given the successes documented to date, future directions will focus on the development and characterization of multichannel reporters in environmentally relevant strains, especially picocyanobacteria from coastal and pelagic environments. The availability of improved culturing protocols for *Prochlorococcus* spp. [122] will facilitate bioreporter construction in the most important genus of subtropical and tropical oligotrophic seas. We also recognize that cyanobacterial reporters provide an incomplete picture of the nutrient status of phytoplankton. Studies assessing aquatic Fe, N, and P bioavailability ideally should be complemented by parallel measurements employing a eukaryotic reporter strain, ideally a diatom. Thus, future efforts should focus equally on developing genetic tools for cyanobacterial and diatom species.

Acknowledgements Cyanobacterial bioreporter development and characterization described in this review was based upon work supported by the National Science Foundation (NSF) under grants OCE-9911592, 0327738 and 0727644 (RMLM and GSB), OCE-0095404 (RMLM) and MCB-9634049 (GSB). The authors recognize the contributions of K. Brinkman, P. Croot, K. Durham, O. Gillor, C. Hassler, N. Ivanikova, C. Pohl, D. Porta, R. Sherrell, R. Sterner, M. Twiss, and S. Wilhelm in furthering our efforts to develop and use nutrient-responsive cyanobacterial bioreporters. We also recognize M. Al-Rshaidat, H. Singler, and T. Villareal for their contributions to measuring variable fluorescence of buoyant diatoms described in the case study presented in Sect. 3.1.1. The authors are grateful to the captains and crews of the R/V *Blue Heron*, R/V *New Horizon* and the CCGS *Limnos* for their assistance in collection of samples.

References

1. Whitton BA, Potts M (2000) Introduction to the cyanobacteria. In: Witton BA, Potts M (eds) The ecology of cyanobacteria: their diversity in time and space. Kluwer, Dordrecht
2. Callieri CJS, Stockner JS (2002) Freshwater autotrophic picoplankton: a review. J Limnol 61:1–14
3. Ting CS, Rocap G, King J, Chisholm SW (2002) Cyanobacterial photosynthesis in the oceans: the origins and significance of divergent light harvesting strategies. Trends Microbiol 10:134–142
4. Goericke R, Welschmeyer NA (1993) The marine prochlorophyte *Prochlorococcus* contributes significantly to phytoplankton biomass and primary production in the Sargasso Sea. Deep Sea Res 40:2283–2294
5. Liu H, Nolla HA, Campbell L (1997) *Prochlorococcus* growth rate and contribution to primary production in the equatorial and subtropical North Pacific Ocean. Aquat Microb Ecol 12:39–47
6. Veldhuis MJW, Kraay GW, Van Bleijswijk JDL, Baars MA (1997) Seasonal and spatial variation in phytoplankton biomass, productivity and growth in the northwestern Indian Ocean: the southwest and northeast monsoon, 1992–1993. Deep Sea Res 44:425–449
7. Fahnenstiel GL, Carrick HJ (1992) Phototrophic picoplankton in Lakes Huron and Michigan: abundance, distribution, composition, and contribution to biomass and production. Can J Fish Aquat Sci 49:379–388
8. Ivanikova NV, Popels L, McKay RML, Bullerjahn GS (2007) Lake Superior supports unique clusters of cyanobacterial picoplankton. Appl Environ Microbiol 73:4055–4066
9. McKay RML, Porta D, Bullerjahn GS, Sterner RW, Brown ET, Sherrell RM (2005) Bioavailable Fe in oligotrophic Lake Superior assessed using biological reporters. J Plankton Res 27:1033–1044

10. Silvert W (1996) Size-aggregation in models of aquatic ecosystems. Sci Total Environ 183:107–114
11. Ware DM, Thompson RE (2005) Bottom-up ecosystem trophic dynamics determine fish production in the Northeast Pacific. Science 308:1280–1284
12. Bachmann T (2003) Transforming cyanobacteria into bioreporters of biological relevance. Trends Biotechnol 21:247–249
13. Belkin S (2003) Microbial whole cell sensing systems of environmental pollutants. Curr Opin Microbiol 6:206–212
14. Andersson CR, Tsinoremas NF, Shelton J, Lebedeva NV, Yarrow J, Min H, Golden SS (2000) Application of bioluminescence to the study of circadian rhythms in cyanobacteria. Methods Enzymol 305:527–42
15. Mackey SR, Ditty JL, Clerico EM, Golden SS (2007) Detection of rhythmic bioluminescence from luciferase reporters in cyanobacteria. In: Rosato E (ed) Circadian rhythms: methods and protocols. Humana, Totowa NJ
16. Kunert A, Hagemann M, Erdmann E (2000) Construction of promoter probe vectors for *Syenechocystis* sp. PCC 6803 using the light-emitting reporter systems Gfp and LuxAB. J Microbiol Methods 41:185–194
17. Boyanapalli R, Bullerjahn GS, Pohl C, Croot PL, Boyd PW, McKay RML (2007) A luminescent whole-cell cyanobacterial bioreporter for measuring Fe availability in diverse marine environments. Appl Environ Microbiol 73:1019–1024
18. Durham KA, Porta D, Twiss M, McKay RML, Bullerjahn GS (2002) Construction and characterization of a luminescent *Synechococcus* sp. PCC7942 Fe-dependent bioreporter. FEMS Microbiol Lett 209:215–221
19. Durham KA, Porta D, McKay RML, Bullerjahn GS (2003) Expression of the iron-responsive *irpA* gene from the cyanobacterium *Synechococcus* sp. strain PCC 7942. Arch Microbiol 179:131–134
20. Gillor O, Hadas O, Post AF, Belkin S (2002) Phosphorus bioavailability monitoring by a bioluminescent sensor strain. J Phycol 38:107–115
21. Gillor O, Harush A, Hadas O, Post AF, Belkin S (2003) A *Synechococcus* P*glnA::luxAB* fusion for estimation of nitrogen bioavailability to freshwater cyanobacteria. Appl Environ Microbiol 69:1465–1474
22. Ivanikova NV, McKay RML, Bullerjahn GS (2005) Construction and characterization of a freshwater cyanobacterial bioreporter capable of assessing nitrate assimilatory capacity in freshwaters. Limnol Oceanogr Methods 3:86–93
23. Mbeunkui F, Richaud C, Etienne AL, Schmid RD, Bachmann TT (2002) Bioavailable nitrate detection in water by an immobilized luminescent cyanobacterial reporter strain. Appl Microbiol Biotechnol 60:306–312
24. Schreiter PP, Gillor O, Post AF, Belkin S, Schmid RD, Bachmann TT (2001) Monitoring of phosphorus bioavailability in water by an immobilized luminescent cyanobacterial reporter strain. Biosens Bioelectron 1:811–818
25. Daunert S, Barrett G, Feliciano JS, Shetty RS, Shrestha S, Smith-Spencer W (2000) Genetically engineered whole-cell sensing systems: coupling biological recognition with reporter genes. Chem Rev 100:2705–2738
26. van der Meer JR, Tropel D, Jaspers M (2004) Illuminating the detection chain of bacterial bioreporters. Environ Microbiol 6:1005–1020
27. Harms H, Wells MC, van der Meer JR (2006) Whole-cell living biosensors-are they ready for environmental application? Appl Microbiol Biotechnol 7:1–8
28. Golden JW, Yoon H-S (2003) Heterocyst development in *Anabaena*. Curr Opinion Microbiol 6:557–563
29. Tolonen AC, Liszt GB, Hess WR (2006) Genetic manipulation of *Prochlorococcus* strain MIT9313: green fluorescent protein expression from an RSF1010 plasmid and Tn5 transposition. Appl Environ Microbiol 72:7607–7613

30. Kitayama Y, Kondo T, Nakahira Y, Nishimura H, Ohmiya Y, Oyama T (2004) An in vivo dual reporter sysem of cyanobacteria using two railroad-worm luciferases with different color emissions. Plant Cell Physiol 45:109–113
31. Boyd TJ, Osburn CL (2004) Changes in CDOM fluorescence from allocthonous and autochthonous sources during tidal mixing and bacterial degradation in two coastal estuaries. Mar Chem 89:189–210
32. Coble PG, Green SA, Blough NV, Gagosian RB (1990) Characterization of dissolved organic matter in the Black Sea by fluorescence spectroscopy. Nature 348:432–434
33. De Souza Sierra MM, Donard OFX, Lamotte M, Belin C, Ewald M (1994) Fluorescence spectroscopy of coastal and marine waters. Mar Chem 47:127–144
34. Kowalczuk P, Stoń-Egiert J, Cooper WJ, Whitehead RF, Durako MJ (2005) Characterization of chromophoric dissolved organic matter in the Baltic Sea by excitation emission matrix fluorescence spectroscopy. Mar Chem 96:273–292
35. Hassler CS, Havens SM, Bullerjahn GS, McKay RML, Twiss MR (2009) An evaluation of iron bioavailability and speciation in western Lake Superior with the use of combined physical, chemical and biological assessment. Limnol Oceanogr 54:987–1001
36. Hakkila K, Maksimow M, Karp M, Virta M (2002) Reporter genes *lucFF, luxCDABE, gfp* and *dsred* have different characteristics in whole-cell bacterial sensors. Anal Biochem 301:235–242
37. Liu Y, Golden SS, Kondo T, Ishiura M, Johnson CH (1995) Bacterial luciferase as a reporter of circadian gene expression in cyanobacteria. J Bacteriol 177:2080–2086
38. Porta D, Bullerjahn GS, Durham KA, Wilhelm SW, Twiss MR, McKay RML (2003) Physiological characterization of a *Synechococcus* sp. (Cyanophyceae) strain PCC7942 iron-dependent bioreporter. J Phycol 39:64–73
39. Shao CY, Howe CJ, Porter AJR, Glover LA (2002) Novel cyanobacterial biosensor for detection of herbicides. Appl Environ Microbiol 68:5026–5033
40. Hastings JW (1978) Bacterial bioluminescence: an overview. Methods Enzymol 57:125–135
41. Hassler CS, Twiss MR, McKay RML, Bullerjahn GS (2006) Optimization of a cyanobacterial (*Synechococcus* sp. PCC 7942) bioreporter to measure bioavailable iron. J Phycol 42:324–335
42. Nodop A, Pietsch D, Höcker R, Becker A, Pistorius EK, Forchhammer K, Michel K-P (2008) Transcript profiling reveals new insights into the acclimation of the mesophilic fresh-water cyanobacterium *Synechococcus elongatus* PCC 7942 to iron starvation. Plant Physiol 147:747–763
43. Yousef N, Pistorius EK, Michel KP (2003) Comparative analysis of *idiA* and *isiA* transcription under iron starvation and oxidative stress in *Synechococcus elongatus* PCC 7942 wild type and selected mutants. Arch Microbiol 180:471–483
44. Hassler CS, Twiss MR, Simon DF, Wilkinson KJ (2008) Porous underwater chamber (PUC) for in-situ determination of nutrient and pollutant bioavailability to microorganisms. Limnol Oceanogr Methods 6:277–287
45. Ivanikova NV, McKay RML, Bullerjahn GS, Sterner RW (2007) Nitrate utilization in Lake Superior is impaired by low nutrient (P, Fe) availability and seasonal light limitation – a cyanobacterial bioreporter study. J Phycol 43:475–484
46. Porta D, Bullerjahn GS, Twiss MR, Wilhelm SW, Poorvin L, McKay RML (2005) Iron bioavailability in Lake Erie (Laurentian Great Lakes) measured by means of a cyanobacterial bioreporter. J. Great Lakes Res 31(Suppl 2):180–194
47. Sakamoto T, Shen G, Higashi S, Murata N, Bryant DA (1997) Alteration of low-temperature susceptibility of the cyanobacterium *Synechococcus* sp. PCC 7002 by genetic manipulation of membrane lipid unsaturation. Arch Microbiol 169:20–28
48. Palenik B, Brahamsha B, Larimer FW, Land M et al (2003) The genome of a motile marine *Synechococcus*. Nature 424:1037–1042
49. Su Z, Dam P, Chen X, Olman V, Jiang T, Palenik B, Xu Y (2003) Computational inference of regulatory pathways in microbes: an application to phosphorus assimilation pathways in *Synechococcus* sp. WH8102. Genome Inform 14:3–13

50. Urbach E, Scanlan DJ, Distel DL, Waterbury JB, Chisholm SW (1998) Rapid diversification of marine picophytoplankton with dissimilar light-harvesting structures inferred from sequences of *Prochlorococcus* and *Synechococcus* (Cyanobacteria). J Mol Evol 46:188–201
51. Brahamsha B (1996) A genetic manipulation system for oceanic cyanobacteria of the genus *Synechococcus*. Appl Environ Microbiol 62:1747–1751
52. Larsen RA, Wilson MM, Guss AM, Metcalf WW (2002) Genetic analysis of pigment biosynthesis in *Xanthomonas autotrophicus* Py2 using a new, highly efficient transposon mutagenesis system that is functional in a wide variety of bacteria. Arch Microbiol 178: 193–201
53. McCarren J, Brahamsha B (2005) Transposon mutagenesis in a marine *Synechococcus* strain: isolation of swimming motility mutants. J Bacteriol 187:4457–4462
54. Palenik B, Ren Q, Dupont CL, Myers GS et al (2006) Genome sequence of *Synechococcus* CC9311: insights into adaptation to a coastal environment. Proc Nat Acad Sci USA 103:13555–13559
55. Boyd PW, Jickells T, Law CS, Blain S et al (2007) Mesoscale iron enrichment experiments 1993–2005: synthesis and future directions. Science 315:612–617
56. Hutchins DA, Di Tullio GR, Zhang Y, Bruland KW (1998) An iron limitation mosaic in the California upwelling regime. Limnol Oceanogr 43:1037–1054
57. Jickells TD, An ZS, Andersen KK, Baker AR et al (2005) Global iron connections between desert dust, ocean biogeochemistry, and climate. Science 308:67–71
58. Hutchins DA, Witter AE, Butler A, Luther GW III (1999) Competition among marine phytoplankton for different chelated iron species. Nature 400:858–861
59. Carpenter SR (1996) Microcosm experiments have limited relevance for community and ecosystem ecology. Ecology 77:677–680
60. Geiß U, Vinnemeier J, Kunert A, Lindner I, Gemmer B, Lorenz M, Hagemann M, Schoor A (2001) Detection of the *isiA* gene across cyanobacterial strains: Potential for probing iron deficiency. Appl Environ Microbiol 67:5247–5253
61. Geiß U, Selig U, Schumann R, Steinbruch R, Bastrop B, Hagemann M, Schoor A (2004) Investigations on cyanobacterial diversity in a shallow estuary (Southern Baltic Sea) including genes relevant to salinity resistance and iron starvation acclimation. Environ Microbiol 6:377–387
62. McKay RML, Wilhelm SW, Hall J, Hutchins DA, Al-Rshaidat MMD, Mioni CE, Pickmere S, Porta D, Boyd PW (2005) The impact of phytoplankton on the biogeochemical cycling of iron in subantarctic waters southeast of New Zealand during FeCycle. Global Biogeochem Cycles 19:GB4S24. doi:10.1029/2005GB002482
63. Xia L, Yakunin AF, McKay RML (2004) The Fe-responsive accumulation of redox proteins ferredoxin and flavodoxin from a marine cryptomonad. Eur J Phycol 39:73–82
64. Webb EA, Moffett JW, Waterbury JB (2001) Iron stress in open-ocean cyanobacteria (*Synechococcus*, *Trichodesmium*, and *Crocosphaera* spp.): identification of the IdiA protein. Appl Environ Microbiol 67:5444–5452
65. Behrenfeld MJ, Sherrell WK, RM CFP, Strutton P, McPhaden M, Shea DM (2006) Controls on tropical Pacific Ocean productivity revealed through nutrient stress diagnostics. Nature 442:1025–1028
66. Hassler CS, Twiss MR (2006) Bioavailability of iron senesed by a phytoplanktonic Fe-bioreporter. Environ Sci Technol 40:2544–2551
67. Michel KP, Pistorius EK (2004) Adaptation of the photosynthetic electron transport chain in cyanobacteria to iron deficiency: the function of IdiA and IsiA. Physiol Plant 120:36–50
68. Singler HR, Villareal TA (2005) Nitrogen inputs into the euphotic zone by vertically migrating *Rhizosolenia* mats. J Plankton Res 27:545–556
69. Villareal TA, McKay RML, Al-Rshaidat MMD, Boyanapalli R, Sherrell RM (2007) Compositional and fluorescence characteristics of the giant diatom *Ethmodiscus* along a 3000 km transect (28°N) in the central North Pacific gyre. Deep-Sea Res I 54:1273–1288

70. Björkman KM, Karl DM (2003) Bioavailability of dissolved organic phosphorus in the euphotic zone at Station ALOHA, North Pacific Subtropical Gyre. Limnol Oceanogr 48:1049–1057
71. Karl DM, Tien G (1992) MAGIC: a sensitive and precise method for measuring dissolved phosphorus in aquatic environments. Limnol Oceanogr 37:105–116
72. Johnson KS, Boyle E, Bruland K, Coale K et al (2007) Developing standard for dissolved iron in seawater. EOS 88:131–132
73. Martin JH, Gordon RM (1988) Northeast Pacific iron distributions in relation to phytoplankton productivity. Deep-Sea Res Part A 35:177–196
74. DeLong EF, Preston CM, Mincer T, Rich V, Hallam SJ, Frigaard N-U, Martinez A, Sullivan MB, Edwards R, Rodriguez Brito B, Chisholm SW, Karl DM (2006) Community genomics among stratified microbial assemblages in the ocean's interior. Science 311:496–503
75. Johnson ZI, Zinser ER, Coe A, McNulty NP, Woodward EMS, Chisholm SW (2006) Niche partitioning among Prochlorococcus ecotypes along ocean-scale environmental gradients. Science 311:1737–1740
76. Geider RJ, La Roche J (1994) The role of iron in phytoplankton photosynthesis, and the potential iron-limitation of primary productivity in the sea. Photosynthesis Res 39:275–301
77. de Baar HJW, de Jong JTM (2001) Distributions, sources and sinks of iron in seawater. In: Turner DR, Hunter KA (eds) The biogeochemistry of iron in seawater. Wiley, Chichester
78. Brown MT, Landing WM, Measures CI (2005) Dissolved and particulate Fe in the western and central North Pacific: Results from the 2002 IOC cruise. Geochem Geophys Geosyst 6 (10). doi:10.1029/2004GC000893
79. Karl DM (2000) Phosphorus, the staff of life. Nature 406:31–32
80. Benitez-Nelson CR (2000) The biogeochemical cycling of phosphorus in marine systems. Earth Sci Rev 51:109–135
81. Karl DM, Björkman KM, Dore JE, Fujieki L, Hebel T, Houlihan T, Letelier RM, Tupas LM (1999) Ecological nitrogen-to-phosphorus stoichiometry at station ALOHA. Deep-Sea Res II 48:1529–1566
82. Karl DM, Yanagi K (1997) Partial characterization of the dissolved organic phosphorus pool in the oligotrophic North Pacific Ocean. Limnol Oceanogr 42:1398–1405
83. Wu J, Sunda W, Boyle EA, Karl DM (2000) Phosphate depletion in the Western North Atlantic Ocean. Science 289:759–762
84. Clark LL, Ingall ED, Benner R (1998) Marine phosphorus is selectively remineralized. Nature 393:426–428
85. Engblom SO (1998) The phosphate sensor. Biosens Bioelectron 13:981–994
86. Boström B, Perrson G, Broberg B (1988) Bioavailability of different phosphorus forms in freshwater systems. Hydrobiologia 170:133–155
87. Hoppe HG (2003) Phosphatase activity in the sea. Hydrobiologia 187:187–200
88. Ammerman JW, Azam F (1985) Bacterial 5'-nucleotidase in aquatic ecosystems: a novel mechanism of phosphorus regeneration. Science 227:1338–1340
89. Dyhrman ST, Webb EA, Anderson DM, Moffett JW, Waterbury JB (2002) Cell-specific detection of phosphorus stress in *Trichodesmium* from the Western North Atlantic. Limnol Oceanogr 47:1832–1836
90. Lomas MW, Swain A, Shelton R, Ammerman JW (2004) Taxonomic variability of phosphorus stress in Sargasso Sea phytoplankton. Limnol Oceanogr 49:2303–2310
91. Fuller NJ, West NJ, Marie D, Yallop M, Rivlin T, Post AF, Scanlan DJ (2005) Dynamics of community structure and phosphate status of picocyanobacterial populations in the Gulf of Aqaba, Red Sea. Limnol Oceanogr 50:363–375
92. Scanlan DJ, Wilson WH (1999) Application of molecular techniques to addressing the role of P as a key effector in marine ecosystems. Hydrobiologia 401:149–175
93. Scanlan DJ, Silman NJ, Donald KM, Wilson WH, Carr NG, Joint I, Mann NH (1997) An immunological approach to detect phosphate stress in populations and single cells of photosynthetic picoplankton. Appl Environ Microbiol 63:2411–2420

94. Dyhrman ST, Haley ST (2006) Phosphorus scavenging in the unicellular marine diazotroph *Crocosphaera watsonii*. Appl Environ Microbiol 72:1452–1458
95. Dyhrman ST, Chappell PD, Haley ST, Moffett JW, Orchard ED, Waterbury JB, Webb EA (2006) Phosphonate utilization by the globally important marine diazotroph *Trichodesmium*. Nature 439:68–71
96. Ilikchyan, I, McKay RML, Zehr JP, Dyhrman ST, Bullerjahn GS (2009) Detection and expression of the phosphonate transporter gene *phnD* in marine and freshwater picocyanobacteria. Environ Microbiol 11:1314–1324
97. Aichi M, Takatani N, Omata T (2001) Role of NtcB in activation of nitrate assimilation genes in the cyanobacterium *Synechocystis* sp. strain PCC 6803. J Bacteriol 183:5840–5847
98. Frias JE, Flores E, Herrero A (2000) Activation of the *Anabaena nir* operon promoter requires both NtcA (CAP family) and NtcB (LysR family) transcription factors. Mol Microbiol 38:613–625
99. Mortonson JA, Brooks AS (1980) Occurrence of a deep nitrite maximum in Lake Michigan. Can J Fish Aquat Sci 37:1025–1027
100. Wilhelm SW, DeBruyn JM, Gillor O, Twiss MR, Livingston K, Bourbonniere RA, Pickell LD, Trick CG, Dean AL, McKay RML (2003) Effect of phosphorus amendments on present day plankton communities in pelagic Lake Erie. Aquat Microb Ecol 32:275–285
101. Sterner RW, Anagnostou E, Brovold S, Bullerjahn GS, Finlay J, Kumar S, McKay RML, Sherrell RM (2007) Increasing stoichiometric imbalance in North America's largest lake: Nitrification in Lake Superior. Geophys Res Letts 34:L10406. doi:10.1029/2006GL028861
102. Koksharova OA, Wolk CP (2002) Genetic tools for cyanobacteria. Appl Microbiol Biotechnol 58:123–137
103. Wolk CP, Cai Y, Panoff J-M (1991) Use of a transposon with luciferase as a reporter to identify environmentally responsive genes in a cyanobacterium. Proc Natl Acad Sci USA 88:5355–5359
104. Maeda S-I, Kawaguchi Y, Ohe T-A, Omata T (1998) *cis*-acting sequences required for NtcB-dependent, nitrite-responsive positive regulation of the nitrate assimilation operon in the cyanobacterium *Synechococcus* sp. Strain PCC 7942. J Bacteriol 180:4080–4088
105. Grossman AR (2005) Paths toward algal genomics. Plant Physiol 137:410–427
106. Walker TL, Collet C, Purton S (2005) Algal transgenics in the genomic era. J Phycol 41:1077–1093
107. Grossman AR, Harris EE, Hauser C, Lefebvre PA et al (2003) *Chlamydomonas reinhardtii* at the crossroads of genomics. Eukaryot Cell 2:1137–1150
108. Ferrante P, Catalanotti C, Bonente G, Giuliano G (2008) An optimized, chemically regulated gene expression system for *Chlamydomonas*. PLoS ONE 3(9):e3200. doi:10.1371/journal.pone.0003200
109. Léon-Bañares R, Gonzáles-Ballester D, Galván A, Fernández E (2004) Transgenic microalgae as green cell-factories. Trends Biotechnol 22:45–52
110. Rosenberg JN, Oyler GA, Wilkinson L, Betenbaugh MJ (2008) A green light for engineered algae: redirecting metabolism to fuel a biotechnology revolution. Curr Op Biotechnol 19:430–436
111. Redinbo MR, Yeates TO, Merchant S (1994) Plastocyanin: structural and functional analysis. J Bioenerg Biomembr 26:49–66
112. Fuhrmann M, Hausherr A, Ferbitz L, Schodl T et al (2004) Monitoring dynamic expression of nuclear genes in *Chlamydomonas reinhardtii* by using a synthetic luciferase reporter gene. Plant Mol Biol 55:869–881
113. Siaut M, Heijde M, Mangogna M, Montsant A, Coesel S, Allen A, Manfredonia A, Falciatore A, Bowler C (2007) Molecular toolbox for studying diatom biology in *Phaeodactylum tricornutum*. Gene 406:23–35
114. Armbrust EV, Berges JA, Bowler C, Green BR, Martinez D et al (2004) The genome of the diatom *Thalassiosira pseudonana*: ecology, evolution, and metabolism. Science 306:79–86

115. Bowler C, Allen AE, Badger JH, Grimwood J et al (2008) The *Phaeodactylum* genome reveals the evolutionary history of diatom genomes. Nature 456:239–244
116. Falciatore A, Ribera d'Alcalà M, Croot P, Bowler C (2000) Perception of environmental signals by a marine diatom. Science 288:2363–2366
117. Poulsen N, Kröger N (2005) A new molecular tool for transgenic diatoms. Control of mRNA and protein biosynthesis by an inducible promoter-terminator cassette. FEBS J 272:3413–3423
118. Poulsen N, Chesley PM, Kröger N (2006) Molecular genetic manipulation of the diatom *Thalassiosira pseudonana* (Bacillariophyceae). J Phycol 42:1059–1065
119. Li J, Xue L, Yan H, Liu H, Liang J (2008) Inducible EGFP expression under the control of the nitrate reductase gene promoter in transgenic *Dunaliella salina*. J Appl Phycol 20:137–145
120. Kröger N (2007) Prescribing diatom morphology: toward genetic engineering of biological nanomaterials. Curr Op Chem Biol 11:662–669
121. Kröger N, Poulsen N (2008) Diatoms – from cell wall biogenesis to nanotechnology. Annu Rev Genet 42:83–107
122. Morris JJ, Kirkegaard R, Szul MJ, Johnson ZI, Zinser ER (2008) Facilitation of robust growth of *Prochlorococcus* colonies and dilute liquid cultures by "helper" heterotrophic bacteria. Appl Environ Microbiol 74:4530–4534

Adv Biochem Engin/Biotechnol 118: 189–210
DOI: 10.1007/10_2009_3

Published online: 21 May 2009

Application of Microbial Bioreporters in Environmental Microbiology and Bioremediation

E. E. Diplock, H. A. Alhadrami, and G. I. Paton

Abstract Bioreporters have been widely acknowledged to represent new and novel approaches in applied microbiology. Despite a plethora of constructions covering a diverse range of detection devices and host organisms, genuine applications are rare. Here, their application in the areas of general environmental microbiology, analytical detection and bioremediation are summarised and critically considered. Future applications require a more integrated approach such that those constructing bioreporters are aware of the needs of the end-user. A decade ago, predictions were made of the pivotal role of bioreporters and our future reliance; this fortune telling may take another decade to reach fruition.

Keywords Bioreporter • Diagnostics • Environmental analysis • Reporter gene • Hydrocarbons • Bioremediation

Contents

E.E. Diplock (✉), H.A. Alhadrami, and G.I. Paton
Institute of Biological and Environmental Sciences, Cruickshank Building, University of Aberdeen, Aberdeen, AB24 3UU, UK
e-mail: e.e.diplock@abdn.ac.uk, h.alhadrami@abdn.ac.uk

E.E. Diplock and G.I. Paton
Remedios Ltd, Aberdeen Science and Technology Park, Aberdeen, AB22 8GW, UK
e-mail: g.i.paton@abdn.ac.uk

1 Introduction

Throughout this book the emphasis has been on the mechanism and performance of bioreporters for a wide range of purposes. These technology advancements reflect what is often perceived as state of the art and their performance and their potential value continues to increase. Genuine applications are rare. Statutory government agencies seem reluctant to adopt them. Surrogate animal testing, risk assessment modelling and chemical analysis still dominate the 'potential market'. Concerns of valid controls and false positives are used by traditionalists to shun their potential.

Nearly two decades after the publication of the first environmental bioreporter paper, a review was published taking on a whole new angle [1]. In many ways it is like a mountaineering journal, the key peaks have been climbed but what has evolved as a consequence, are the techniques and the apparatus to make ascents quicker, to tackle more difficult routes and to respond to new challenges. Van der Meer [1] succinctly summarises this when there is the consideration of environmental analytical chemists realising that bioreporters offer a new frontier by addressing different questions with alternative tools; complementing or even extending the value of current data assimilation.

Bioreporters by their definition are responsive promoters fused to living organisms and as such they offer tremendous advantages in their application. To address this point through the use of examples, the research areas have been partitioned into discrete areas; the detection and tracking of target organisms and the analysis of chemicals of concern and/or their toxicity burden. In each one of these disciplines the bioreporter offers a genuine advantage over a direct culturable or molecular technique or against chemical analysis, but often the research is published by relating the virtues of bioreporter systems relative to the traditional alternative approach thus 'selling short' the technology.

2 Applications of Bioreporters in the Detection of Target Organisms

In terms of this research area, the two most significant applications have been in the study of microorganisms in the context of human pathogens and their role in plant soil interactions. Conventional methods for detection and identification of environmental pathogens are adequately sensitive, cost effective, and provide quantitative and qualitative information of the number and the nature of the tested microorganisms [2]. Such approaches make use of morphological evaluation of the microorganisms, biochemical screening, serological confirmation, selective enrichment, or assessment of the ability of microorganisms to grow in various types of defined media under specific conditions [3]. It is, however, acknowledged that certain bacterial strains in certain environmental conditions may become viable but non-culturable, rendering cultivation approaches of little value [4]. Bioreporter approaches enable the unequivocal detection of isolates introduced into complex matrices and have been applied to terrestrial and marine systems, bioprocesses, and foodstuffs [2, 3]. A summary of key recent applications is shown in Table 1. These examples have been considered in terms of reporter gene, host strain and application.

The potential application of bioluminescent bioreporters for the detection of bacteria was originally through the development of luciferase reporter phages by Ulitzur and Kuhn [15]. The genes encoding luciferase into the genome of bacteriophage enabled a phenotypic response by host bacteria to infection. Reporter phages enable detection of bacterial host cells following phage infection with a relatively short incubation time [16]. The reported level of detection using such a technology is as low as 10 *E. coli* cells, 100 *S. typhimurium* cells, and 10 enterobacterial cells [16]. Sarkis et al. [17] used the L5 bacteriophage to detect *M. segmantis*. It was possible to detect 100 cells of *M. segmantis* in a few hours and 10 cells in 2 days by this bacteriophage approach [17]. Different *Salmonella* species and *Listeria* were also detected by employing the same approach [18]. *L. monocytogenes* is an opportunistic food-borne pathogen which causes human listeriosis via ingestion of

Table 1 Applications of bioreporters for monitoring environmental microorganisms

Reporter genes	Host strains	Applications	References
luxI	Bacteriophage *PP01-luxI*	Detection several strains of *E. coli* O157:H7	[5]
luxCDABE, *luxI* and *luxR*	*luxI incorporated reporter lambda phage* and *E. coli OHHLux*	General detection of *E. coli* in pure culture, and other pathogenic agents at low population densities. In addition, it has been employed for the specific sensing of the auto-inducer *N*-(3-oxo-hexanoyl$_{)\text{-L-}}$homoserine lactone	[6]
gfp	*P. syringae*	Monitoring of several plants pathogens	[7]
gfp	*S. mutans*	Monitoring of biofilm formation	[8]
gfp	*E. coli*	Specifically tagged gram-negative bacteria like *E. coli* and *P. aeruginosa*, as well as other Gram-negative bacteria in contaminated water sampled from the environment	[9]
gfp	*P. syringae* and *E. coli*	Measurement the availability of water to a bacterial cell	[10]
lacZ	*P. fluorescens*	Monitoring of low oxygen habitats in soil	[11].
gfp	Bacteriophage *P22*	Monitoring bacterial exposure to transient or low levels of specific molecules directly in the environment	[12]
dmpR-LuxAB	*P. putida KT2440*	Has been used for determining biodegradation properties in soil	[13]
luxCDABE	*P. putida RB1353*	Analytical tool for in situ determination of bioavailable levels of contaminants in static environmental samples, and an indicator of metabolic activity in bacterial cells in response to changes in environmental conditions such as oxygen tension	[14]

contaminated food [16]. Conventional methods for detecting the genus *Listeria* includes enrichment and plating on selective media, following by biochemical identification and species conformation. Reporter bacteriophage A511::*luxAB* was constructed by Loessner et al. [16] for detection of a wide range of *Listeria* strains. They reported recombinant phage A511::*luxAB* as viable and infective through mediation of luciferase expression in the infected cells. This detection (and the correlation of luminescence and cell numbers) was possible when there were large numbers of other microorganisms present. As a test system, when the user is able to make use of introduced cells, this system is cost effective, rapid, and simple. However, there are a few weaknesses; namely false-negative responses that can occur, particularly from *L. monocytogenes*. Furthermore, isolation and species differentiation might be essential when using this system as members of other *Listeria* species (e.g. *L. ivanovii*, *L. seeligeri*, *L. innocua* and *L. welshimeri*) are prone to infection by A511 [16]. Sensitive and specific bioreporter bacteriophage were developed by Blasco et al. [19] for *S. newport* and *E. coli*. Sensitivity improvement was achieved by focusing on the bacterial adenylate kinase as the cell marker instead of ATP. Cell number was proportional to light emission, and 103 cells were detectable in a sample volume of 0.1 mL [19].

A variety of bioreporter bacteriophage-based approaches have been used for monitoring and detection of environmental microorganisms [20–27]. Some of these approaches require addition of external substrates or use of sophisticated equipment which limits their applications [29]. To overcome some of the limitations associated with the previous bacteriophage approaches, a two-component bacteriophage-based bioluminescent reporter system for the detection of *E. coli* in environmental samples has been developed [30]. The system consists of a *luxI* incorporated into a lambda bacteriophage and a *lux*-based bioluminescent reporter cell, which responds to the infection event through acyl-homoserine lactone (AHL) mediated quorum sensing and bioluminescent signal stimulation. *E. coli* O157:H7 is a particularly aggressive human pathogen with exposure pathways that include soil, water and direct food ingestion [31]. Cattle and other domestic animals are natural reservoirs of *E. coli* O157:H7 [22]. *E. coli* O157:H7 has been shown to enter the viable but non-culturable state following exposure to oligotrophic fresh and sea waters at an ambient temperature [28]. This unequivocal evidence was possible with a bioreporter approach. The virulent phage (PP01) has the ability to infect *E. coli* O157:H7 strains with high specificity [32]. This phage was used by Ode et al. [28] for the rapid and sensitive detection of *E. coli* O157:H7. Further studies, such as the assignment of a positive vs negative cut-off point, are necessary for the application of this tool in water and food enrichment samples [28].

Such bioreporter approaches have also been used in rhizosphere investigations with success [33]. Fundamentally though these approaches require the destructive introduction of the bioreporters into environmental samples and as such they fail to report in genuine pristine environments. Their application can enhance our understanding of processes and indeed complement models, but they fundamentally require the disruption of environmental samples.

3 Use of Bioreporters in Environmental Analysis and Diagnostics

Bioreporters for the detection of specific elements or biodegradation pathways have been constructed by the fusion of appropriate promoters to reporter genes [34]. The challenge however is to develop a viable bioassay that enables the analysis of the sample in a form that is as near to real environmental conditions as possible. As such, the procedure should be performed at the appropriate environmental pH and under similar osmotic conditions, with the addition of no substrates or complexing agents. A suitable extraction procedure and the ability to analyse a sample with little cross contamination or competing 'inducing agents', is essential.

To apply the bioreporters to environmental samples there are several different strategies. The environmental samples may be exposed to a constitutively marked bioreporter and the response may be deemed to be indicative of a given pollution burden. The bioreporter could also be constructed such that a specific analyte induces the reporter gene. Finally, a bioreporter could be constructed such that it responds to materials that pose mutagenic or carcinogenic modes of action. Most reviews consider each of these three constructions differently, but in this review the analyte is acknowledged to drive the selection of the given bioreporter. Girotti et al. [34] in a comprehensive appraisal consider ecotoxicity and then the component role of the sensor technology. While this is the most routine way to critically appraise literature, the problem is that it lacks a level of integration. This means that the end-users are often unfamiliar with some of the considerable constraints that certain pollutants impose upon the role and value of such technology.

It is essential that the nature of the bioassay employed must be taken into consideration when selecting a bioreporter, as assay procedures have been reported to affect sensor sensitivity and therefore performance [35]. Inducible sensors are often constructed with the aim of detecting specific pollutants, whereas other isolates are designed for the determination of general toxicity and it is therefore of interest in determining their response to environmental samples. In particular, where environmental samples are concerned, it is highly unlikely that a given site would be contaminated with a single type of pollutant; in fact, it is far more likely that a site would be co-contaminated with a number of pollutants [36].

Biological responses to given chemicals of concern are not random. Quantitative structure activity relationships (QSAR), which relate the response to physicochemical attributes have been the focus of many studies, mainly for organic compounds [37] but also for inorganic analytes [38]. There are a limited number of modes of actions exhibited by given analytes and once this has been determined for a given receptor and chemical form, biological response can be correlated against a physicochemical parameter (e.g. partitioning coefficient, Henry's Law constant). Furthermore, Trott et al. [37] revealed that the derived QSAR for a microbial bioreporter could be correlated with the response of a higher organism (invertebrate or indeed vertebrate) if the mode of action of the chemical was the same. Often made overly complex by authors, these QSAR are valuable because they enable toxicity prediction for unknown chemicals and they also serve to validate the actual assay as it is applied. Furthermore, their value is not solely with assessing toxicity but can be used when considering the susceptibility of given analytes to induce specific genes as Bundy et al. [39] demonstrated. This approach was further evolved when a suite of QSAR was established to a range of induction-based bioreporters that included responses to toluene, octane, isopropyl benzene and naphthalene and their appropriate analogues [40]. Such approaches have, as yet, not been established in elemental responses because resistance genes have greater specificity. Also, the ability to plot a range of elements, that are similar, yet have a variable in a given physicochemical constant is very limited.

So these remarks cover the basis of the evolution of the assay and the validation of the procedures and represent the bridge between bioreporter design and application.

Table 2 is a summary of the genuine environmental applications of bioreporters. The use of naturally luminescent systems has been excluded as this would make the list unmanageable and would do little to complement the preceding sections of the book.

Hydrocarbon contamination is noted to be ubiquitous, but this is not reflected in the use of reporters. In part this is because refined products pose little harm to microbial receptors and such materials are far from ideal in achieving effective induction. Dawson et al. [41] reported that using an appropriate extract, bioreporters had value but only when seen in the context of other soil ecotoxicological assays. Seifonova and Eaton [43] and Bundy et al. [42] reported that the most useful value for the bioreporters, when applied to solvent extracted soil samples, was when the interpretation was done in combination with speciated chemical analysis. Of greater success and more widely applied are studies that focus on particular hydrocarbons. The BTEX compounds have received attention both from toxicity perspectives and the induction of degradation genes. Willardson et al. [46], Stinter and Halverson [45] and Dawson et al. [44] were able to apply bioreporters for the analysis of environmental samples directly (wastewaters) or through simple aqueous extraction techniques (soils and sediments). The relatively high water solubility highlighted the merit of induction based approaches, which correlated well with chemical analysis. Phenols are more polar than BTEX and have received attention as potential pollutants by Leedjarv et al. [48] who reported positive results in the laboratory, but poor performance for environmental samples. If these problems could be resolved, there is considerable scope for such systems being applicable to natural processes where phenolic compounds may be a key indicator of land degradation. Leedjarv et al. [48] made use of the induction response of a catabolic pathway and were able to study the response of substituted phenols. Tiensing et al. [49] managed to use constitutive bioreporters in soils for monitoring chlorophenols and found close correlation with chemical analysis. Moreover, they also considered the interactive effects of combined chlorophenol doses using simple additive models.

Bhattacharyya et al. [50] studied the extent and impact of a chlorinated solvent plume on an aquifer. This approach effectively made use both of constitutive and degradation induced bioreporters. The results of the bioreporters needs to be interpreted in the context of detailed chemical analysis and there was a need to make sure that appropriate negative and positive controls were adopted for assay calibration.

Table 2 Applications of bioreporters for environmental diagnostics and analysis

Analytes	Reporter genes	Host strains	Applications	Conclusion	References
Hydrocarbons	*luxCDABE*	*P. fluorescens 10586r, E. coli HB101* and *E. coli CM1166*	Ecotoxicological comparative evaluation of soils undergoing a range of remediative strategies: 26 different assays were considered	No single test was well suited to this complex analysis but bioreporters in conjunction with other approaches had value	[41]
Hydrocarbons	*luxCDABE*	*P. fluorescens 10586r, E. coli HB101* and *E. coli CM1166*	Changes in hydrocarbon bioavailability and toxicity during remediation	More useful when coupled with chemical analysis	[42]
Hydrocarbons	*Ibp-luxCDABE*	*E. coli HMS174*	Monitoring a mixture of hydrocarbons such as jet fuel JP4 and creosote and ethanol extracts of contaminated sediment from Chesapeake Bay	Complements chemical analysis by evolving biological assessments	[43]
BTEX (benzene, toluene, ethylbenzene, and xylenes)	*luxCDABE*	*P. fluorescens 10586r* and *P. putida TVA8*	Monitoring the bioremediation of a mixture of BTEX compounds in a trial	The combined use of bioreporters enabled an informative insight into the response and this complemented other microbial assays and chemical approaches	[44]
BTEX (benzene, toluene, ethylbenzene, and xylenes)	*gfp*	*P. fluorescens*	Detection of toluene and benzene derivatives compounds	Combined approaches of biological and chemical are the best option	[45]
BTEX (benzene, toluene, ethylbenzene, and xylenes)	*xylR-luc*	*E. coli DH5α*	Monitoring BTEX compounds in aquifer water and contaminated soil samples	Complementary tool for chemical analysis	[46]
Phenols	*luxCDABE*	*Acinetobacter DF4*	Assessment the ecotoxicity of water samples contaminated with phenols	More appropriate for monitoring the influent toxicity than Microtox assay	[47]
Phenols	*luxCDABE*	*P. fluorescens OS8*	Detection of phenolic compounds from oil-shale industry pollution as well as phenol contaminated leachate and groundwater	It is useful for the general estimation of the bioavailability of phenolic pollution. However, measuring of individual compounds in the natural samples was impossible	[48]

(continued)

Table 2 (continued)

Analytes	Reporter genes	Host strains	Applications	Conclusion	References
Chlorophenols	*luxCDABE*	*P. fluorescens 10586r* and *E. coli HB101*	Toxicity measurement in soil water extract	More informative that chemical analysis in isolation	[49]
Chlorinated solvents	*luxCDABE*	*P. putida TVA8, P. Fluorescens 10586, E. coli HB101* and *E. coli DH5α*	Bioavailability assessment of chlorinated solvents emanating from a plume into an aquifer	Complements chemical analysis by biological assessments	[50]
Organotins	*luxCDABE*	*P. fluorescens 10586r, E. coli HB101* and *V. Fischeri*	To monitor changes in toxicity during the remediation and degradation of phenyltins in soil	Useful application and added meaning to the cumbersome process of the chemical analysis	[51]
Range of metals	*luxCDABE*	*Acinetobacter DF4*	Monitoring the ecotoxicity of water, wastewater and effluent samples contaminated with heavy metals such as Zn, Cd, Fe, Co, Cr and Cu	Useful for general toxicity monitoring of water and wastewater samples but further chemical analysis for the toxic sample is required	[52]
Range of metals	*luxCDABE*	*P. fluorescens 10586r, E. coli HB101* and *R. leguminosarum*	To compare the binding response of a wide range of contrasting soils and to enable a biological interpretation of solution chemistry	Correlates with measured solution concentration	[53]
Range of metals	*luxCDABE*	*E. coli HB101*	Toxicity measurement in soils and comparison with other sentinel soil ecotoxicity assessments	Bioreporters correlate closely with solution chemistry and may offer an early warning of longer term microbial impacts	[54]
Range of metals	*luxCDABE*	*P. fluorescens Shk1*	Toxicity monitoring for activated sludge system exposed to heavy metals (Cu, Zn, Ni, and Cd) in a bench-scale wastewater treatment system. Shk1 bioluminescence was most sensitive to Cd and Zn, followed by Cu and Ni	Valuable for effluent monitoring but, an assay for protecting activated sludge microorganisms should by developed to mimic the accumulated effect of toxicants via adsorption	[55]

Copper (Cu)	*gfpuv*	*S. cerevisiae DY150*	Detection of copper ions in soil	Complementary tool for chemical analysis	[56]
Copper and nickel (Cu and Ni)	*luxCDABE*	*P. fluorescens 10586r, E. coli HB101* and *A. Mellea*	Comparative soil toxicity testing in smelter impacted areas	Bioreporters correlated closely with solution chemistry measurements. A relatively sensitive response when compared to other microbial assays	[57]
Copper and arsenic (Cu and As)	*luxCDABE*	*P. fluorescens 10586r* and *E. coli HB101 E. coli CM1166*	Measurement of arsenate, arsenite and copper toxicity and bioavailability in sediment and water	Combined bioreporters yield useful results when interpreted alongside chemical analysis	[58]
Arsenic (As)	*luxCDABE*	*E. coli DH5α*	Toxicity assessment of rice consumption by measuring inorganic arsenic extracted from rice, and differentiation between the organic and inorganic forms of arsenic in different rice samples	Despite a long assays, the procedure negated the need for chemical testing and allowed high throughput	[59]
Arsenic (As)	*luxAB*	*E. coli DH5α*	Large scale quantification of arsenic in natural groundwater samples with reduction of the potential disturbances generated by high iron	Rapid screening tool for As in developing countries, and can be upgraded to multiwell-plate analyses	[60]
Antimony (Sb)	*luxCDABE*	*P. fluorescens 10586r* and *E. coli HB101 E. coli CM1166*	Bioavailability and toxicity measurements of antimony and other metals and metalloids in soils	Combined approaches of biological and chemical are the best option	[61]
Mercury and cadmium (Hg and Cd)	*mer-lux* and	*E. coli HMS174*	The impact of pH on the uptake and intracellular accumulation of Hg(II) and Cd(II) by *E. coli* under aerobic conditions	Enhances our understanding of bacterial accumulation of Hg under anaerobic conditions and the impact of pH	[62]
Iron (Fe)	*luxAB*	*Synechococcus PCC 7002 Fe*	Assessment of Fe availability in different marine environments, and Fe availability to phytoplankton	Should be considered as a complementary tool for chemical	[63]

(continued)

Table 2 (continued)

Analytes	Reporter genes	Host strains	Applications	Conclusion	References
Iron (Fe)	*luxAB*	*Cyanobacteria*	Assessment the bioavailability of Fe in Lake	Influence of Fe generated by the added reporter cells while using unfiltered water samples should be taken into account	[64]
Iron (Fe)	*luxAB genes*	*Synechococcus sp. PCC 7942*	Serves as a quantitative tool to assess Fe deficiency in natural freshwater environments	Complementary tool for chemical analysis	[65]
Nitrate (NO_3^-)	*InZ* and *gfp*	*E. coli DH5α* and *E. cloacae EcCT501R*	Measurement of availability of nitrate in rhizosphere, and microbial niches in natural environments	Highly responsive to nitrate in culture and in the rhizosphere. The two reporters were complementary to each other in terms of nitrate estimation	[33]
Effluents contamination	*lacZ*	*E. coli DH5α*	Assessment of the ecological status of a small river basin in Belgium contaminated with industrial and wastewater effluents	More complex river basins required to be assessed before a conclusion about a more general applicability can be made	[66]
Estrogenic compounds	*luxCDABE* and *laZ*	*S. cerevisiae BLYES* and *the Yeast laZ reporter strain (YES)*	Screening, chemical sensing and real time monitoring of estrogenic compounds and endocrine-disrupting chemicals in the environment	Should be combined with other technology, such as photodetection, for screening a large number of estrogenic compounds. Yeast-based assays are unable to differentiate between estrogen agonists and antagonists	[67]
N-(3-Oxo-hexanoyl)-l-homoserine lactone	*luxR* and *luxCDABE*	*E. coli ROlux2*	Quantification the quorum sensing in *V. fischeri* by measuring the signalling molecule *N*-(3-oxo-hexanoyl)-l-homoserine lactone	Quantitative tool, dose not require extraction of the signalling molecules and can be used with small sample volume	[68]

Regulatory proteins HbpR and XylR	*luxAB*	*E. coli DH5α*, *P. azelaica* and *P. putida PaW164*	Monitoring regulatory proteins activate specific degradation. Furthermore, a dual-responsive *P. azelaica* was constructed, containing both HbpR and XylR and stimulating luciferase expression from the same signal promoter independently with for 2-hydroxybiphenyl and *m*-xylene	A powerful tool as a dual- responsive bioreporter	[69]
Phenylacetic acid (PAA) ($C_8H_8O_2$)	*gfp*	*P. putida KT2440*	Detection of the bacterial antifungal agent phenylacetic acid (PAA)	Useful tool for detecting bacteria producing PAA	[70]
Ethanol and sodium chloride (C_2H_6O and NaCl)	*GFPuv gene*	*E. coli*	Monitoring of physiological status of cells with respect to the growth phase and estimation of ethanol and NaCl as two environmental stress factors	The activity of *rpoS* promoter which was used was an effective device for evaluating the cellular physiological status	[71]
Oxytetracycline ($C_{22}H_{24}N_2O_9$)	*gfp*	*E. coli*	Detection of oxytetracycline production by *S. rimosus* in soil	Valuable for monitoring the production of oxytetracycline through *S. rimosus*	[72]
Sucrose and tryptophan ($C_{12}H_{22}O_{11}$ and $C_{11}H_{12}N_2O_2$)	*InZ*	*E. herbicola 299RTice* and *E. herbicola 299R(p61RYice)*	Detection of sucrose and tryptophan in rhizosphere	Additional bioreporters should be developed to detect the other components of root exudates, as this one is useful to detect only a fraction of the microbially relevant components of root exudation	[73]

The majority of bioreporter applications have been in the study of metals and metalloids in the environment. Metal resistance does not always confer a high level of specificity [74–76]. For example, reporters designed to respond to both Cd and Pb have been used to determine the bioavailable fraction of the metals in environmental samples [77, 78]. In a sense, the constitutively marked bioreporters offer less ambiguous responses than those that respond to specific analytes, because there is a straightforward dose response relationship. Two whole-cell bacterial bioreporters (*S. flexneri pLux* and *S. sonnei pLux*) were constructed by Olaniran et al. [79] and applied as a wastewater biodegradation monitoring system. Advantages reported included that they were cost effective, able to cope with large sample runs and are capable of generating a consistent pattern suitable for use as an indicator for biodegradation. Ultimately, it was proposed that these bioreporters could be applied for the on-line monitoring of wastewater treatment plants [79].

Abd-El-Haleem et al. [52], Dawson et al. [53], Kelly et al. [55] and Paton et al. [51] all reported clear does relationships between measured metal concentrations and the expression of bioluminescence. Most importantly, different host organisms had differential responses and the assays were applicable to complex matrices including soil, slurries and wastewater. Tandy et al. [54] and Paton et al. [57] evaluated the response of bioreporters against other microbial and higher organism assays. Overall, the finding was that there was a close correlation with the measured chemical parameters and in general the bioreporter was more sensitive than other assays. This result had also been reported by Dawson et al. [53] who described the responses by means of EC_{10} values when considering the partitioning behaviour of metals in soils.

A panel of indicator strains containing fusions of different stress promoters to *luxCDABE* was evaluated for detecting antibacterial agents at low doses and to predict the mode of action of the tested compounds from their bioluminescence response [80]. An ability to predict the mode of action could have significant value in both environmental and pharmaceutical applications in the future. Findings revealed that different antibacterial modes of action generate specific responses and that discrete signatures could be compiled.

The role of bioreporters in predicting concentration has been fundamental to many researchers working with specific elements. Porta et al. [64], Durham et al. [65] and Boyanapalli et al. [63] used bioreporters for the measurement of iron in natural environments and to understand the biological processing of this key element. The detection was observed to correlate with iron deficiencies in natural systems, in particular the great lakes. While this may be a valuable tool in terms of aqueous samples, it is likely to endure interference when applied to soils or sediments. Mercury has been the focus of most attention when constructing bioreporters and, while their sensitivity is high in aqueous samples, the performance in soil, sediments and wastewater is much poorer, mainly due to false positives and interference from other ions. *E. coli* RBE23-17 was constructed using a *zntA*-based transcriptional switch, and was used to quantify concentrations of Cd in seawater and soil samples [81]. A number of arsenate and arsenite sensors have also been developed [82–84, 106], including an As-specific sensor that was reported to be applicable to environmental conditions [85]. Cu-inducible bioreporters have also been successfully applied in the monitoring of Cu-contaminated soils [86, 87].

One of the key weaknesses of an uninformed application of such bioreporters is that the nature of the response of the cells cannot be matched to chemical concentration. This is because the initial peak in luminescence declines as the toxicity impact takes effect and as such that a given measure of light needs to be calibrated against a particular location in a nonlinear response. To circumvent this issue, Flynn et al. [58, 61] took a combined use of both As resistance marked and constitutive bioreceptors and reported a positive response when there was some knowledge of all the competing ions present in a complex sample. More recently, there has been an evolution of techniques that are able to operate with more complex matrices. The key constraint is that, just because these are bioreporters, it does not mean that they measure bioavailability. Indeed, bioavailability as a term offers great controversy, but it must be the case that the value of the bioreporter is greatest when coupled to a relevant extraction technique. Golding et al. [62] developed an assay capable of operating under a range of relevant redox conditions, while Trang et al. [60] and Baumann and Van Der Meer [59] used As responsive bioreporters, under clever, environmentally relevant extraction approaches, to relate responses to higher organism toxicity. In particular the innovation of using bioreporters to measure As concentration in human urine represents a genuine innovative approach in inorganic analysis. Patterson et al. [88] considered the value of XAD resin and cyclodextrin when applied to assessing the bioavailability of hydrophobic compounds and were able to connect the response with biodegradation effectiveness. As a consequence the key issue worthy of consideration, relates to what the actual meaning of bioluminescence is in the case of these examples. Is it a measure of concentration or of a biologically defined component? To a large extent the literature remains split on this issue.

However, in the case of analysis of nutrients, it is perhaps more clear-cut that instead of carrying out a saline step (typically KCl), a bioreporter can interrogate directly an environmental sample for the form and concentration of nutrients [33]. The nitrogen deficiency promoter *glnA*, is induced with various forms of N while the *phoA* promoter responds to P [89]. The value of these approaches is that the opportunity for application is significant, limited only by the constraint of the ability to measure and to ensure resolution. In a similar theme, signalling molecules which may prove particularly difficult to analyse are also worthy of inclusion [68, 107]. Jaeger et al. [73] refined extraction procedures for the interrogation and quantification of sucrose and tryptophan in the rhizosphere. There are an increasing number of research groups making use of such tools, but more as a proof of concept rather than enhancing fundamental knowledge.

The value of QSAR has already been considered. Most compounds are described as narcotics (polar and non-polar). To a large extent though, these are not the current key analytes of interest. Detection and quantification of analytes has moved on to represent compounds that belong to Class IV. These molecules have specific modes of action and are mainly of concern for human toxicity. The most significant example is the detection and quantification of estrogenic and estrogene mimicking molecules.

Such bioreporters have been widely reported in construction and more recently in environmental applications [67]. Bioreporters have key advantages in that they prove much more cost effective than existing analytical approaches and are able to group similar molecules together; that means a quantification of the receptor response rather than individual analysis. While a great deal of attention has been given to aqueous samples, the number of environmental applications is rising, and the adoption of suitable assays sympathetic to low water solubility, are becoming more widely adopted.

4 Bioreporter Application in the Discipline of Bioremediation

Early bioreporter construction was done specifically under the guise of the needs of bioremediation [90, 91]. The main justification of using bioreporters for bioremediation is that they allocate effective assessment of physiochemical constraints to bioremediation and thus allow optimisation of hydrocarbon processing by the alleviation of such constrains [92, 93]. Furthermore, they are regarded as a cost effective tool for site remediation as they offer the opportunity of on-line as well as off-line monitoring of contaminants [92, 94]. Bioreporters are claimed to indicate the bioavailable fraction of complex mixtures of contaminants found in soil [74]. Two *lux*-marked bioluminescent bacteria were used by [95] to assess the progress of bioremediation for soil contaminated with oil. It was reported that *P. putida* F1 (pUCD607) was more sensitive to paraffin than to motor oil. However, this is not a measure of remediation potential. Indeed it was concluded that the use of community profile techniques, in parallel with the toxicity bioassay could develop a better insight of the toxic impacts of oil contamination and can help to assess oil bioremediation [95].

Some researchers (e.g. [50, 93]) have successfully used constitutive bioreporters to interrogate environmental samples that have been put through 'extraction stages', referred to as sample manipulation. These stages have developed to reflect the type of options that a remediation practitioner may require to carry out in the field. Analysing samples that are contaminated with volatile aromatic and chlorinated compounds, those with an adverse pH, the occurrence of non-volatile organic contaminants and the presence of heavy metals, could all be defined and mitigations applied. Such approaches were well suited to groundwater but found to be less applicable to soils and sediments.

Natural attenuation of petroleum hydrocarbons, which relies on in situ bioremediation, has received substantial attention over time [96]. As the bioremediation of these sites is based upon the presence of microorganisms which are capable of biodegrading BTEX compounds, there is a need to confirm that the environmental status is appropriate for attenuation.

The *tod-luxCDABE* fusion was constructed by Applegate et al. [96], and introduced into the chromosome of *P. putida* F1, yielding the strain *P. putida* TVA8. Consequently, this bioreporter has been applied for BTEX compounds and chlorinated solvents.

This bioreporter has also been used by Shingleton et al. [97] to assess the bioavailability of BTEX compounds in aqueous extracts, initially by developing a standard dose-response curve to determine a maximum luminescence concentration and then subsequently in a biodegradation experiment [97]. The bioreporter was further used by Dawson et al. [44] to monitor the degradation of BTEX compounds and their related toxicity as biodegradation proceeds in soils undergoing three different remedial treatments.

Ripp et al. [98] carried out an investigation to test whether *P. fluorescens* HK44 could be effectively introduced and maintained in a soil environment, and to assess the applications of in situ bioluminescence as a tool for monitoring and control of the online bioremediation. The results were positive and enabled an insight into resilience and performance, highlighting the value of a marker-based reporter system [98].

5 All These Developments but Where Next?

Bioreporter measurements can add an important dimension to the prediction of fate and availability of particular analytes in the environment. At a more applied level, such bioreporters could transform pollutant monitoring and predictive bioremediation [29]. Bioreporters need to be engineered to address some of the limitations of this technology [29, 92]. For instance, most published bacterial bioreporters are not sensitive to pollutant concentrations below 0.1 μM, except notably Hg. Consequently, further optimisation of the bioreporters performance in terms of sensitivity, speed of the response, the choice of targets to detect, and analyte detection limit should be influenced and improved [29]. In addition, we need 'complete sensor engineering', if we are truly serious about the future of microbial bioreporters [29].

Bioreporters have the ability to detect groups of compounds rather than single analytes, which can be a useful feature when individual chemicals of a group have similar toxicity or bioremediative properties [29]. Furthermore, they have a high quality potential for equipment miniaturisation due to their small size [99]. Such bioreporters also offer a genuine tool in fundamental science, as the constructions today are revealed to be unsurpassed in studying gene expression and the physiology of bacteria in complex environments [100].

Girotti et al. [34] make a comprehensive appraisal of the techniques available for measuring responses but do not consider genuine integration or miniaturisation. Bioluminescent bioreporter integrated circuits (BBIC) in which marked organisms are integrated with integrated circuits enables sensing and reporting of low concentrations of varieties of toxic substances in both gas and liquid environments [101]. The circuits are basically a microluminometer consisting of photodiodes. The photodiode emission is transferred into a digital signal and the proportion and frequency is proportional to the concentration of the pollutants. This system was tested to detect naphthalene in gaseous phase and salicylate in liquid phase, and it showed great ability for sensing very low concentrations of environmental pollutants.

Currently, bioreporters need to interface with target molecules either following a suitable organic solvent extraction step or directly via the gas phase. Werlen et al. [102] pioneered a technique that was refined and adopted by various authors including Tecon et al. [103] and Kohlmeier et al. [104] which shows great promise and sensitivity, but has not yet been applied to soil samples. When making use of organic solvents, bioreporters must be exposed to them after appropriate dilution. This approach is problematic for two reasons: (1) the use of organic solvents often has negative adverse effects on bioreporters and (2) the portion of the HOC that partitions into the organic solvent extractions does not reflect HOC bioavailability, thus undermining the very principle of using bioreporters when applied to soil and sediment samples.

6 Concluding Remarks

Bioreporters may offer an excellent solution for some environmental challenges in the future. In particular, recent work that has made use of novel and very relevant extraction techniques and in terms of the developing world, low cost analytical devices may offer in field testing. Too much energy has been spent on trying to compete on a level playing field with analytical chemical techniques, when fundamentally the procedures should be complementary. The approach is becoming more integrated because detection devices are being considered in parallel with the bioengineering. As confidence in the bioreporter, assay and detection device increase, the adoption will be evolutionary. Bioreporters are excellent tools for measuring microscale environments [105] and this is where the future challenges in resolution will be. A similarity was drawn between mountaineering and bioreporters and this is now more pertinent than ever because the importance in both disciplines is about security, confidence, adopting technology and meeting new challenges. Popularity will determine the level of confidence and extent of this adoption.

References

1. Van Der Meer JR (2006) Analytics with engineered bacterial bioreporter strains and systems. Curr Opin Biotechnol 17:1–3
2. Leonard P, Hearty S, Brennan J et al. (2003) Advances in biosensors for detection of pathogens in food and water. Enzyme Microb Technol 32:3–13
3. Ivnitski D, Abdel-Hamid I, Atanasov P et al. (1999) Biosensors for detection of pathogenic bacteria. Biosens Bioelectron 14:599–624
4. Toze S (1999) PCR and the detection of microbial pathogens in water and wastewater. Water Res 33:3545–3556
5. Brigati JR, Ripp SA, Johnson CM et al. (2007) Bacteriophage-based bioluminescent bioreporter for the detection of *Escherichia coli* O157:H7. J Food Prot 70:1386–1392
6. Ripp S, Jegier P, Birmele M et al. (2006) Linking bacteriophage infection to quorum sensing signalling and bioluminescent bioreporter monitoring for direct detection of bacterial agents. J Appl Microbiol 100:488–499

7. Sabaratnam S, Beattie GA (2003) Differences between *Pseudomonas syringae* pv. syringae B728a and *Pantoea agglomerans* BRT98 in epiphytic and endophytic colonization of leaves. Appl Environ Microbiol 69:1220–1228
8. Rice AR, Hamilton MA, Camper AK (2003) Movement, replication, and emigration rates of individual bacteria in a biofilm. Microb Ecol 45:163–172
9. Goh YY, Ho B, Ding JL (2002) A novel fluorescent protein-based biosensor for gram-negative bacteria. Appl Environ Microbiol 68:6343–6352
10. Axtell CA, Beattie GA (2002) Construction and characterization of a *proU-gfp* transcriptional fusion that measures water availability in a microbial habitat. Appl Environ Microbiol 68:4604–4612
11. Højberg O, Schnider U, Winteler HV et al. (1999) Oxygen-sensing reporter strain of *Pseudomonas fluorescens* for monitoring the distribution of low-oxygen habitats in soil. Appl Environ Microbiol 65:4085–4093
12. Casavant NC, Beattie GA, Phillips GJ et al. (2002) Site-specific recombination-based genetic system for reporting transient or low-level gene expression. Appl Environ Microbiol 68: 3588–3596
13. Sarand I, Skärfstad E, Forsman M et al. (2001) Role of the DmpR-mediated regulatory circuit in bacterial biodegradation properties in methylphenol-amended soils. Appl Environ Microbiol 67:162–171
14. Neilson JW, Pierce SA, Maier RM (1999) Factors influencing expression of *luxCDABE* and nah genes in *Pseudomonas putida* RB1353(NAH7, pUTK9) in dynamic systems. Appl Environ Microbiol 65:3473–3482
15. Ulitzur S, Kuhn J (1987) Introduction of lux genes into bacteria, a new approach for specific determination of bacteria and their antibiotic susceptibility. In: Sclomerich J, Andreesen R, Kapp A, Ernst M, Woods WG (eds) Bioluminescence and chemiluminescence: new perspectives. Wiley, Bristol, UK, pp 463–472
16. Loessner MJ, Rees CED, Stewart GSAB et al. (1996) Construction of luciferase reporter bacteriophage A511:*luxAB* for rapid and sensitive detection of viable *Listeria* cells. Appl Environ Microbiol 62:1133–1140
17. Sarkis GJ, Jacobs WR Jr, Hatfull GF (1995) L5 luciferase reporter mycobacteriophages: a sensitive tool for the detection and assay of live mycobacteria. Mol Microbiol 15:1055–1067
18. Chen J, Griffiths MW (1996) *Salmonella* detection in eggs using *lux* + bacteriophages. J Food Prot 59:908–914
19. Blasco R, Murphy MJ, Sanders MF et al. (1998) Specific assays for bacteria using phage mediated release of adenylate kinase. J Appl Microbiol 84:661–666
20. Pagotto F, Brovko L, Griffiths MW (1996) Phage-mediated detection of *Staphylococcus aureus* and *Escherichia coli* using bioluminescence. Bacteriol Qual Raw Milk 9601:152–156
21. Goodridge L, Chen J, Griffiths M (1999) Development and characterization of a fluorescent-bacteriophage assay for detection of *Escherichia coli* O157:H7. Appl Environ Microbiol 65:1397–1404
22. Waddell TE, Poppe C (2000) Construction of mini-Tn*10luxABcam/Ptac*-ATS and its use for developing a bacteriophage that transduces bioluminescence to *Escherichia coli* O157:H7. FEMS Microbiol Lett 182:285–289
23. Favrin SJ, Jassim SA, Griffiths MW (2001) Development and optimization of a novel immunomagnetic separation-bacteriophage assay for detection of *Salmonella enterica* serovar enteritidis in broth. Appl Environ Microbiol 67:217–224
24. Goodridge L, Griffiths M (2002) Reporter bacteriophage assays as a means to detect foodborne pathogenic bacteria. Food Res Int 35:863–870
25. Funatsu T, Taniyama T, Tajima T et al. (2002) Rapid and sensitive detection method of a bacterium by using a GFP reporter phage. Microbiol Immunol 46:365–369
26. Mosier-Boss PA, Lieberman SH, Andrews JM et al. (2003) Use of fluorescently labeled phage in the detection and identification of bacterial species. Appl Spectrosc 57:1138–1144
27. Casavant NC, Thompson D, Beattie GA et al. (2003) Use of a site-specific recombination-based biosensor for detecting bioavailable toluene and related compounds on roots. Environ Microbiol 5:238–249

28. Oda M, Morita M, Unno H et al. (2004) Rapid detection of *Escherichia coli* O157:H7 by using green fluorescent protein-labeled PP01 bacteriophage. Appl Environ Microbiol 70:527–534
29. Van Der Meer JR, Tropel D, Jaspers M (2004) Illuminating the detection chain of bacterial bioreporters. Environ Microbiol 6:1005–1020
30. Birmele M, Ripp S, Jegier P et al. (2008) Characterization and validation of a bioluminescent bioreporter for the direct detection of *Escherichia coli*. J Microbiol Methods 75:354–356
31. Ackman D, Marks S, Mack P et al. (1997) Swimming-associated haemorrhagic colitis due to *Escherichia coli* O157:H7 infection: evidence of prolonged contamination of a fresh water lake. Epidemiol Infect 119:1–8
32. Morita M, Fischer CR, Mizoguchi K et al. (2002) Amino acid alterations in Gp38 of host range mutants of PP01 and evidence for their infection of an ompC null mutant of *Escherichia coli* O157:H7. FEMS Microbiol Lett 216:243–248
33. DeAngelis KM, Ji P, Firestone MK et al. (2005) Two novel bacterial biosensors for detection of nitrate availability in the rhizosphere. Appl Environ Microbiol 71:8537–8547
34. Girotti S, Ferri EN, Fumo MG et al. (2008) Monitoring of environmental pollutants by bioluminescent bacteria. Anal Chim Acta 608:2–29
35. Bierkens J, Klein G, Corbisier P et al. (1998) Comparative sensitivity of 20 bioassays for soil quality. Chemosphere 37:2935–2947
36. Farreé M, Arranz F, Riboó J et al. (2004) Inter-laboratory study of the bioluminescence inhibition tests for rapid wastewater toxicity assessment. Talanta 62:549–558
37. Trott D, Dawson JJC, Killham KS et al. (2007) Comparative evaluation of a bioluminescent bacterial assay in terrestrial ecotoxicity testing. J Environ Monitor 9:44–50
38. McKinney JD, Richard A, Waller C et al. (2000) The practice of structure activity relationships (SAR) in toxicology. Toxicol Sci 56:8–17
39. Bundy JG, Morriss AWJ, Durham DG et al. (2001) Development of QSARs to investigate the bacterial toxicity and biotransformation potential of aromatic heterocylic compounds. Chemosphere 42:885–892
40. Paton GI, Bundy JG, Campbell CD et al. (2004) Application of catabolic based biosensors to develop QSARs for degradation. In: Cronin MTD, Livingstone DJ (eds) Predicting chemical toxicity and fate, 1st edn. CRC, Boca Raton
41. Dawson JJC, Godsiffe EJ, Thompson IP et al. (2007) Application of biological indicators to assess recovery of hydrocarbon impacted soils. Soil Biol Biochem 39:164–177
42. Bundy JG, Campbell CD, Paton GI (2001) Comparison of response of six different luminescent bacterial bioassays to bioremediation of five contrasting oils. J Environ Monitor 3:404–410
43. Selifonova OV, Eaton RW (1996) Use of an *ipb-lux* fusion to study regulation of the isopropylbenzene catabolism operon of *Pseudomonas putida* RE204 and to detect hydrophobic pollutants in the environment. Appl Environ Microbiol 62:778–783
44. Dawson JJC, Iroegbu CO, Maciel H et al. (2008) Application of luminescent biosensors for monitoring the degradation and toxicity of BTEX compounds in soils. J Appl Microbiol 104:141–151
45. Stiner L, Halverson LJ (2002) Development and characterization of a green fluorescent protein-based bacterial biosensor for bioavailable toluene and related compounds. Appl Environ Microbiol 68:1962–1971
46. Willardson BM, Wilkins JF, Rand TA et al. (1998) Development and testing of a bacterial biosensor for toluene-based environmental contaminants. Appl Environ Microbiol 64:1006–1012
47. Zaki S, Abd-El-Haleem D, Abulhamd A et al. (2008) Influence of phenolics on the sensitivity of free and immobilized bioluminescent *Acinetobacter* bacterium. Microbiol Res 163:277–285
48. Leedjärv A, Ivask A, Virta M et al. (2006) Analysis of bioavailable phenols from natural samples by recombinant luminescent bacterial sensors. Chemosphere 64:1910–1919
49. Tiensing T, Strachan N, Paton GI (2002) Evaluation of interactive toxicity of chlorophenols in water and soil using *lux*-marked biosensors. J Environ Monitor 4:482–489
50. Bhattacharyya J, Read D, Amos S et al. (2005) Biosensor-based diagnostics of contaminated groundwater: assessment and remediation strategy. Environ Pollut 134:485–492
51. Paton GI, Cheewasedtham W, Marr IL et al. (2006) Degradation and toxicity of phenyltin compounds in soil. Environ Pollut 144:746–751

52. Abd-El-Haleem D, Zaki S, Abulhamd A et al. (2006) *Acinetobacter* bioreporter assessing heavy metals toxicity. J Basic Microbiol 46:339–347
53. Dawson JJC, Campbell CD, Towers W et al. (2006) Linking biosensor responses to Cd, Cu and Zn partitioning in soils. Environ Pollut 142:493–500
54. Tandy S, Barbosa V, Tye A et al. (2005) Comparison of different microbial bioassays to assess metal-contaminated soils. Environ Toxicol Chem 24:530–536
55. Kelly CJ, Tumsaroj N, Lajoie CA (2004) Assessing wastewater metal toxicity with bacterial bioluminescence in a bench-scale wastewater treatment system. Water Res 38:423–431
56. Shetty RS, Deo SK, Liu Y et al. (2004) Fluorescence-based sensing system for copper using genetically engineered living yeast cells. Biotechnol Bioeng 88:664–670
57. Paton GI, Viventsova E, Kumpene J et al. (2006) An ecotoxicity assessment of contaminated forest soils from the Kola Peninsula. Sci Total Environ 355:106–117
58. Flynn HC, Mc Mahon V, Diaz GC et al. (2002) Assessment of bioavailable arsenic and copper in soils and sediments from the Antofagasta region of northern Chile. Sci Total Environ 286:51–59
59. Baumann B, Van Der Meer JR (2007) Analysis of bioavailable arsenic in rice with whole cell living bioreporter bacteria. J Agr Food Chem 55:2115–2120
60. Trang PTK, Berg M, Viet PH et al. (2005) Bacterial bioassay for rapid and accurate analysis of arsenic in highly variable groundwater samples. Environ Sci Technol 39:7625–7630
61. Flynn HC, Meharg AA, Bowyer PK et al. (2003) Antimony bioavailability in mine soils. Environ Pollut 124:93–100
62. Golding GR, Sparling R, Kelly CA (2008) Effect of pH on intracellular accumulation of trace concentrations of Hg(II) in *Escherichia coli* under anaerobic conditions, as measured using a mer-*lux* bioreporter. Appl Environ Microbiol 74:667–675
63. Boyanapalli R, Bullerjahn GS, Pohl C et al. (2007) Luminescent whole-cell cyanobacterial bioreporter for measuring Fe availability in diverse marine environments. Appl Environ Microbiol 73:1019–1024
64. Porta D, Bullerjahn GS, Twiss MR et al. (2005) Determination of bioavailable Fe in lake Erie using a luminescent cyanobacterial bioreporter. J Gt Lakes Res 31:180–194
65. Durham KA, Porta D, Twiss MR et al. (2002) Construction and initial characterization of a luminescent *Synechococcus sp.* PCC 7942 Fe-dependent bioreporter. FEMS Microbiol Lett 209:215–221
66. Dardenne F, Smolders R, De Coen W et al. (2007) Prokaryotic gene profiling assays to detect sediment toxicity: evaluating the ecotoxicological relevance of a cell-based assay. Environ Sci Technol 41:1790–1796
67. Sanseverino J, Gupta RK, Layton AC et al. (2005) Use of *Saccharomyces cerevisiae* BLYES expressing bacterial bioluminescence for rapid, sensitive detection of estrogenic compounds. Appl Environ Microbiol 71:4455–4460
68. Yan L, Allen MS, Simpson ML et al. (2007) Direct quantification of *N*-(3-oxo-hexanoyl)-l-homoserine lactone in culture supernatant using a whole-cell bioreporter. J Microbiol Methods 68:40–45
69. Tropel D, Baähler A, Globig K et al. (2004) Design of new promoters and of a dual-bioreporter based on cross-activition by the two regulatory proteins *XylR* and HbpR. Environ Microbiol 6:1186–1196
70. Kim J, Jeon CO, Park W (2007) A green fluorescent protein-based whole-cell bioreporter for the detection of phenylacetic acid. J Microbiol Biotechnol 17:1727–1732
71. Funabashi H, Haruyama T, Mie M et al. (2002) Non-destructive monitoring of *rpoS* promoter activity as stress marker for evaluating cellular physiological status. J Biotechnol 95:85–93
72. Hansen LH, Ferrari B, Sørensen AH et al. (2001) Detection of oxytetracycline production by *Streptomyces rimosus* in soil microcosms by combining whole-cell biosensors and flow cytometry. Appl Environ Microbiol 67:239–244
73. Jaeger CH, Lindow SE, Miller W et al. (1999) Mapping of sugar and amino acid availability in soil around roots with bacterial sensors of sucrose and tryptophan. Appl Environ Microbiol 65:2685–2690
74. Ramanathan S, Ensor M, Daunert S (1997) Bacterial biosensors for monitoring toxic metals. Trends Biotechnol 15:500–506

75. Nies DH (1999) Microbial heavy-metal resistance. Appl Microbiol Biotechnol 51:730–750
76. Borremans B, Hobman JL, Provoost A et al. (2001) Cloning and functional analysis of the pbr lead resistance determinant of *Ralstonia metallidurans* CH34. J Bacteriol 183:5651–5658
77. Tauriainen S, Karp M, Chang W et al. (1998) Luminescent bacterial sensor for cadmium and lead. Biosens Bioelectron 13:931–938
78. Ivask A, François M, Kahru A et al. (2004) Recombinant luminescent bacterial sensors for the measurement of bioavailability of cadmium and lead in soils polluted by metal smelters. Chemosphere 55:147–156
79. Olaniran AO, Motebejane RM, Pillay B (2008) Bacterial biosensors for rapid and effective monitoring of biodegradation of organic pollutants in wastewater effluents. J Environ Monitor 10:889–893
80. Eltzov E, Ben-Yosef DZ, Kushmaro A et al. (2008) Detection of sub-inhibitory antibiotic concentrations via luminescent sensing bacteria and prediction of their mode of action. Sens Actuators B 129:685–692
81. Biran I, Babai R, Levcov K et al. (2000) Online and in situ monitoring of environmental pollutants: electrochemical biosensing of cadmium. Environ Microbiol 2:285–290
82. Corbisier P, Ji G, Nuyts G et al. (1993) *LuxAB* gene fusions with the arsenic and cadmium resistance operons of *Staphylococcus aureus* plasmid pI258. FEMS Microbiol Lett 110:231–238
83. Tauriainen S, Karp M, Chang W et al. (1997) Recombinant luminescent bacteria for measuring bioavailable arsenite and antimonite. Appl Environ Microbiol 63:4456–4461
84. Ramanathan S, Shi W, Rosen BP et al. (1998) Bacteria-based chemiluminescence sensing system using β-galactosidase under the control of the ArsR regulatory protein of the ars operon. Anal Chim Acta 369:189–195
85. Roberto FF, Barnes JM, Bruhn DF (2002) Evaluation of a GFP reporter gene construct for environmental arsenic detection. Talanta 58:181–188
86. Tom-Petersen A, Hosbond H, Nybroe O (2001) Identification of copper-induced genes in *Pseudomonas fluorescens* and use of a reporter strain to monitor bioavailable copper in soil. FEMS Microbiol Ecol 38:59–67
87. Stoyanov JV, Magnani D, Solioz M (2003) Measurement of cytoplasmic copper, silver, and gold with a lux biosensor shows copper and silver, but not gold, efflux by the CopA ATPase of *Escherichia coli*. FEBS Lett 546:391–394
88. Patterson CJ, Semple KT, Paton GI (2004) Non-exhaustive extraction techniques (NEETs) for the prediction of naphthalene mineralisation in soil. FEMS Microbiol Lett 241:215–220
89. Purohit HJ, (2003) Biosensors as molecular tools for use in bioremediation. J Clean Prod 11:293–301
90. King JMH, Di Grazia PM, Applegate B et al. (1990) Rapid, sensitive bioluminescent reporter technology for naphthalene exposure and biodegradation. Science 249:778–781
91. Harms H, Wells MC, Van Der Meer JR (2006) Whole-cell living biosensors – are they ready for environmental application?Appl Microbiol Biotechnol 70:273–280
92. Bailey M, Corbisier P, Glover LA et al. (1998) Reporter genes for monitoring microbial cell activity and/or the environment. An opinion. EU Biotechnol Programme DGXII
93. Sousa S, Duffy C, Weitz H et al. (1998) Use of a *lux*-modified bacterial biosensor to identify constraints to bioremediation of BTEX-contaminated sites. Environ Toxicol Chem 17:1039–1045
94. Van der Lelie D, Corbisier P, Baeyens W et al. (1994) The use of biosensors for environmental monitoring. Res Microbiol 145:67–74
95. Bundy JG, Paton GI, Campbell CD (2004) Combined microbial community level and single species biosensor responses to monitor recovery of oil polluted soil. Soil Biol Biochem 36:1149–1159
96. Applegate BM, Kehrmeyer SR, Sayler GS et al. (1998) A chromosomally based *tod-luxCDABE* whole-cell reporter for benzene, toluene, ethybenzene, and xylene (BTEX) sensing. Appl Environ Microbiol 64:2730–2735
97. Shingleton JT, Applegate BM, Nagel AC et al. (1998) Induction of the tod operon by trichloroethylene in *Pseudomonas putida* TVA8. Appl Environ Microbiol 64:5049–5052

98. Ripp S, Nivens DE, Ahn Y et al. (2000) Controlled field release of a bioluminescent genetically engineered microorganism for bioremediation process monitoring and control. Environ Sci Technol 34:846–853
99. Nivens DE, McKnight TE, Moser SA et al. (2004) Bioluminescent bioreporter integrated circuits: potentially small, rugged and inexpensive whole-cell biosensors for remote environmental monitoring. J Appl Microbiol 96:33–46
100. Hansen LH, Sørensen SJ (2001) The use of whole-cell biosensors to detect and quantify compounds or conditions affecting biological systems. Microb Ecol 42:483–494
101. Vijayaraghavan R, Islam SK, Zhang M et al. (2007) A bioreporter bioluminescent integrated circuit for very low-level chemical sensing in both gas and liquid environments. Sens Actuators B 123:922–928
102. Werlen C, Jaspers MCM, Van Der Meer JR (2004) Measurement of biologically available naphthalene in gas and aqueous phases by use of a *Pseudomonas putida* biosensor. Appl Environ Microbiol 70:43–51
103. Tecon R, Wells M, Van Der Meer JR (2006) A new green fluorescent protein-based bacterial biosensor for analysing phenanthrene fluxes. Environ Microbiol 8:697–708
104. Kohlmeier S, Mancuso M, Tecon R et al. (2007) Bioreporters: gfp versus *lux* revisited and single-cell response. Biosens Bioelectron 22:1578–1585
105. Tecon R, Van Der Meer JR (2006) Information from single-cell bacterial biosensors: what is it good for?Curr Opin Biotechnol 17:4–10
106. Ji and Silver (1992) Regulation and expression of the Arsenic Resistance Operon from Staphylococcus aureus Plasmid Pi258. 1992. Journal of Bacteriology, 174, 3684–3694
107. DeAngelis KM, Firestone MK and Lindow SE (2007) Sensitive whole-cell biosensor suitable for detecting a variety of *N*-Acyl homoserine lactones in intact rhizosphere microbial communities. Applied and Environmental Microbiology, 73, 3724–3727.

Index